Edited by
Soroush Nazarpour and Stephen R. Waite

Graphene Technology

Edited by Soroush Nazarpour and Stephen R. Waite

Graphene Technology

From Laboratory to Fabrication

WILEY-VCH
Verlag GmbH & Co. KGaA

Editors

Dr. Soroush Nazarpour
Group NanoXplore Inc.
25 Montpellier Blvd
Montreal
QC H4N 3K7
Canada

Stephen R. Waite
Graphene Stakeholders Association
640 Ellicott Street, Suite 499
Buffalo
NY 14203
United States

Library of Congress Card No.: applied for

British Library Cataloguing-in-Publication Data
A catalogue record for this book is available from the British Library.

Bibliographic information published by the Deutsche Nationalbibliothek
The Deutsche Nationalbibliothek lists this publication in the Deutsche Nationalbibliografie; detailed bibliographic data are available on the Internet at <http://dnb.d-nb.de>.

Print ISBN: 978-3-527-33833-7
ePDF ISBN: 978-3-527-68757-2
ePub ISBN: 978-3-527-68755-8
Mobi ISBN: 978-3-527-68756-5
oBook ISBN: 978-3-527-68754-1

Cover Design Formgeber, Mannheim, Germany
Typesetting SPi Global, Chennai, India
Printing and Binding Markono Print Media Pte Ltd, Singapore

Printed on acid-free paper

Contents

List of Contributors

Sarbajit Banerjee
Texas A&M University
Department of Chemistry
College Station
TX 77842-3012
USA

and

Texas A&M University
Department of Materials Science
and Engineering
575 Ross Street
College Station
TX 77843-3003
USA

Suman Chhetri
Surface Engineering & Tribology
Division
Council of Scientific and
Industrial Research-Central
Mechanical Engineering
Research Institute
Durgapur 713209
India

and

Academy of Scientific and
Innovative Research (AcSIR)
CSIR-CMERI, Campus
Durgapur 713209
India

Rachel D. Davidson
Texas A&M University
Department of Chemistry
580 Ross Street
College Station
TX 77842-3012
USA

and

Texas A&M University
Department of Materials Science
and Engineering
575 Ross Street
College Station
TX 77843-3003
USA

Robert V. Dennis
Texas A&M University
Department of Chemistry
580 Ross Street
College Station
TX 77842-3012
USA

and

Texas A&M University
Department of Materials Science
and Engineering
575 Ross Street
College Station
TX 77843-3003
USA

Siegfried Eigler
Friedrich-Alexander-Universität
Erlangen-Nürnberg (FAU)
Central Institute of Materials and
Processes and
Department of Chemistry and
Pharmacy
Dr.-Mack-Str. 81
D-90762 Fürth
Germany

and

Chalmers University of
Technology
Department of Chemistry and
Chemical Engineering
Kemivägen 10
SE-412 96 Göteborg
Sweden

José L. Figueiredo
Laboratory of Separation and
Reaction Engineering
Laboratory of Catalysis and
Materials (LSRE-LCM)
Chemical Engineering
Department
Faculdade de Engenharia
Universidade do Porto
Rua Dr. Roberto Frias
4200-465 Porto
Portugal

Nathan A. Fleer
Texas A&M University
Department of Chemistry
580 Ross Street
College Station
TX 77842-3012
USA

and

Texas A&M University
Department of Materials Science
and Engineering
575 Ross Street
College Station
TX 77843-3003
USA

Tapas Kuila
Surface Engineering & Tribology Division
Council of Scientific and Industrial Research-Central Mechanical Engineering Research Institute
Durgapur 713209
India

and

Academy of Scientific and Innovative Research (AcSIR)
CSIR-CMERI, Campus
Durgapur 713209
India

Sergio Morales-Torres
Laboratory of Separation and Reaction Engineering
Laboratory of Catalysis and Materials (LSRE-LCM)
Chemical Engineering Department
Faculdade de Engenharia
Universidade do Porto
Rua Dr. Roberto Frias
4200-465 Porto
Portugal

Naresh Chandra Murmu
Surface Engineering & Tribology Division
Council of Scientific and Industrial Research-Central Mechanical Engineering Research Institute
Durgapur 713209
India

and

Academy of Scientific and Innovative Research (AcSIR)
CSIR-CMERI, Campus
Durgapur 713209
India

Soroush Nazarpour
Group NanoXplore Inc.
25 Montpellier Blvd
Montreal
QC H4N 3K7
Canada

Luisa M. Pastrana-Martínez
Laboratory of Separation and Reaction Engineering
Laboratory of Catalysis and Materials (LSRE-LCM)
Chemical Engineering Department
Faculdade de Engenharia
Universidade do Porto
Rua Dr. Roberto Frias
4200-465 Porto
Portugal

Adrián M.T. Silva
Laboratory of Separation and Reaction Engineering
Laboratory of Catalysis and Materials (LSRE-LCM)
Chemical Engineering Department
Faculdade de Engenharia
Universidade do Porto
Rua Dr. Roberto Frias
4200-465 Porto
Portugal

Marko Spasenovic
Graphene Tracker
Center for Solid State Physics
and New Materials
Institute of Physics
Pregrevica 118
11030 Belgrad
Serbia

Stephen R. Waite
Graphene Stakeholders
Association
640 Ellicott Street, Suite 499
Buffalo
New York 14203
USA

Cyrus Zamani
University of Tehran
School of Metallurgy and
Materials Engineering
College of Engineering
University of Tehran
North Kargar Street
P. O. Box 14395-515
Tehran
Iran

1
Graphene Technology: The Nanomaterials Road Ahead

Stephen R. Waite and Soroush Nazarpour

A new paradigm is emerging for advanced nanomaterials and their use in commercial products. We call it "molecular precision manufacturing" (MPM), and it is evolving as a consequence of the need to develop new tools, new standards, new protocols, and new processes (TSPPs) to foster the commercialization of nanomaterials.

Nanomaterials possess extraordinary properties, but harnessing these properties for use in commercial products is challenging. The emerging MPM paradigm is required in order to realize the tremendous commercial potential of advanced nanomaterials – both 2D and 3D – discovered over the past 25 years.

The TSPPs associated with MPM have been in development for several decades. They combine activities that are critical to the use of advanced nanomaterials in products and applications: 2D materials, such as graphene, molybdenum disulfide, and boron nitride; and 3D nanomaterials, such as single-wall and multi-wall carbon nanotubes (CNTs). Additionally, technologies have been developed to functionalize these advanced 2D and 3D nanomaterials to enhance their properties for use in commercial products. We are at an early stage in the evolution of functionalized advanced 2D and 3D nanomaterials, but the research done thus far is encouraging.

1.1
Newly Discovered 2D Materials

The past decade has witnessed the discovery of several 2D nanomaterials, all of which possess unique properties suited to various applications. These discoveries include the following:

Graphene: Single layer of carbon atoms only 1 molecule thick packed in a hexagonal lattice.

Molybdenum disulfide (MoS_2): When stacked, MoS_2 looks and feels like graphite. However, it is very different from graphene at the 2D level. While graphene is a flat layer of carbon atoms, MoS_2 is composed of molybdenum atoms sandwiched between two sulfur atoms. Unlike graphene, in its natural

Graphene Technology: From Laboratory to Fabrication,
First Edition. Edited by Soroush Nazarpour and Stephen R. Waite.

form it can serve as a semiconductor in transistors, making it appealing for use in electronics and solar cells. Scientists have been experimenting with combining the two materials to allow graphene to have transistor-friendly properties, but are now looking at using MoS_2 on its own. It has properties similar to silicon, but requires the use of much less material and consumes less energy.

Silicene: When silicon is reduced to a 1-atom-thick layer, it takes on a slightly squished-looking honeycomb structure similar to graphene. Like molybdenum disulfide, it can be used as a transistor in its natural form. Silicene also shares one of graphene's especially interesting properties: electrons move through it at a very fast pace, as if they were massless. This means that silicene conducts electricity faster than any commercially available semiconductor. Because silicon is so ubiquitous in current electronics, silicene could be much easier to adopt than other 2D materials. It was only recently synthesized for the first time last year, so the research will take some time to mature. It also could turn out to be more difficult to make than graphene.

Germanane: The element germanium has already been used as a semiconductor, and actually formed the very first transistors in the 1940s. When reduced to a single layer of atoms, it forms a material known as germanane. Germanane conducts electrons 5 times faster than germanium and 10 times faster than silicon, which makes it ideal for creating faster computer chips. It is more stable than silicon and a better absorber and emitter of light. Manufacturers may also be able to produce it on existing equipment in large quantities, which would give it an advantage over emerging graphene manufacturing techniques.

Our experience of working with 2D nanomaterials is limited, given their relatively recent discovery – in the case of graphene, as recent as 2004. Working with 2D materials presents a set of learning curves that require scaling even before the potential of such promising materials can be realized. The TSPPs associated with the emerging MPM paradigm are critical to the commercialization of products and applications using 2D nanomaterials and their 3D counterparts.

Commercialization demands that one has a consistent and repeatable product available at a rational price, given the performance impact and value proposition. Creating the strongest composite in the world is of no value if its mechanical properties cannot be predicted or relied upon because of inconsistent materials or testing. Without these TSPPs, we are not likely to see the fruits anticipated with nanotechnology that many analysts have envisioned, given its vast potential in commercial applications.

In the following text, we offer an overview of MPM and shed light on the promises and challenges associated with the emerging MPM paradigm.

1.2 Wonder Materials

The ascent of MPM is associated with the discovery of "bulk" nanomaterials possessing remarkable properties. We make the distinction between bulk materials and nanoscale elements of electronic and semiconductor devices, for example, which are created as sub-micron architectures using processes such as chemical vapor deposition and epitaxial growth, but which are not "freestanding" materials.

One of the early nanomaterial discoveries came from Rice University in the mid 1980s, with the synthesis of fullerenes, commonly referred to as *bucky-balls* – hollow, spherical carbon structures that became an early impetus to research in novel carbon allotropes. The discovery led to more investigation in Japan on hollow tubes of carbon in the early 1990s and ignited great interest in single- and multi-wall CNTs. CNTs were seen to have a host of remarkable properties that stimulated the interest of nanotechnology researchers all over the world, and it was not long before patent filings on CNT-based applications began to skyrocket.

In 2004, researchers Andre Geim and Kostantin Novoselov from the University of Manchester discovered graphene – another nanomaterial possessing truly extraordinary properties. In 2010, Geim and Novoselov were awarded the Nobel Prize in Physics for their discovery of this "wonder material," which comprises a single layer of carbon atoms only 1 molecule thick (hence its 2D classification) and packed in a hexagonal lattice. It is the thinnest material known to man, with an exceptionally high theoretical surface area (2630 $m^2 g^{-1}$). Atomically, it is the strongest material ever measured, is extremely elastic (stretchable), and has exceptional thermal and electrical conductivity, making it the substance a design engineer's dreams.

Understandably, graphene-related patent filings have risen significantly around the world over the past several years. The United Kingdom is currently a hotbed of activity in graphene, with the University of Manchester acting as a magnet for millions of dollars of research funding. In 2013, the European Union created a Flagship to promote the development of graphene, committing 1 billion Euros in funding over a 10-year time frame. Entrepreneurial activity and investment associated with graphene has increased significantly. Technology stalwarts Samsung and IBM have been extremely active in patenting graphene-based applications. The Far East has been massively active not only in patent applications, but also in investment. Singapore, for example, boasts the highest level of graphene research funding as a percentage of GDP in the world.

With the discovery of graphene in 2004, we have entered a new age of materials and materials science. Since then, several other 1-atom or 1-molecule-thick crystals have been isolated and tentatively studied. These materials range from semiconducting monolayers to wide-gap insulators to metals. This growing library of 2D materials opens the potential to construct various 3D structures with on-demand properties that do not occur naturally, but can be assembled "Lego-style" by stacking individual atomic planes on top of one another in a desired sequence (see Section 1.1).

The discovery of new advanced nanomaterials – both 2D and 3D – over the past 25 years has generated much excitement and hype, which is understandable in light of their remarkable properties. Today, the range of potential applications for graphene and other 2D materials is limited only by one's imagination. Yet, this potential needs to be tempered by the kind of level-headedness that comes from experience working with advanced nanomaterials.

In May 2013, Bayer Material Science (BMS) exited the CNT business and shuttered its production plant, after many years of work and millions' worth of investment. BMS CEO Patrick Thomas noted that while the company remains convinced that CNTs have huge potential (they initially talked of over 3000 tons of output), their experience suggests that potential areas of application that once seemed promising from a technical standpoint are currently either extremely fragmented or do not overlap with the company's core products and spectrum of applications.

At the time of exiting the business, it was reported that BMS had invested some $30 million to produce multi-walled CNTs with a facility that had a capacity of producing over 200 tons per year. Mitsubishi Corp. had a similar experience in the 1990s when it attempted to scale and commercialize fullerenes. While no public information has been made available, insiders indicated that as much as $60 million was invested and, to date, no commercial products realized.

While sobering, the BMS experience holds many valuable lessons for those seeking to commercialize advanced nanomaterials. The commercialization of advanced nanomaterials, and nanotechnology in general, is unlike anything ever undertaken before. Successful commercialization of these advanced nanomaterials requires new approaches, tools, and processes, and a great deal of what seems to be in short supply these days with investors: patience. Often, to satisfy the demands of investors, substantial claims are made on production volumes and estimated sales prior to evaluating the market and without exercising caution.

Arriving at a pure material virtually free from the catalysts used in the production process was not as easy as expected. The challenge was compounded by the need to functionalize these materials; to aid dispersion, acids were often used (as that was all that was available then). High levels of functionalization required a vicious circle of excessive acid treatment, with higher resultant costs, waste streams, and structural degradation.

Crucially, the effect of nanomaterials on the target medium is often not known or precisely predicted until it is attempted. Experience shows that taking a process from the lab (micro) level to the commercialization (macro) level is not easy, and in scaling up, the results can often be different from the lab-based results. This will affect commercial outcomes, possibly rendering a positive projected return to an uneconomic position. It is here that we encounter the classic case of overpromising and under-delivery, effectively stunting the market.

Having to learn these important lessons the hard way is common in business – through failed multimillion-dollar investments, layoffs, plant sales, and closure. Yet, it would be foolhardy to extrapolate failures associated with the development of CNTs into the future, for the very success with advanced nanomaterials lies in these failures. Thomas Edison, Nikola Tesla, and Steve Jobs are just a few famous examples of innovators whose failures led to successes beyond their wildest dreams. Fostering a culture of acceptance of failure as a learning process that moves one closer to success is crucial. How often is a failure seen as unacceptable, resulting in management changes that may not be justified?

Failure is instructive, and a large part of the innovation process. That said, it is important to respect the meme that insanity is doing the same thing over and over and expecting different results, as Einstein once observed. What we learn from the failures of working with advanced nanomaterials is that traditional approaches and processes do not work, and something else is required. This is where MPM comes in.

1.3 The Rise of MPM

Humans have been figuring out how to turn various materials into useful products since the Stone Age. While some of this knowledge scales into the commercialization of nanomaterials today, new learning curves are clearly required to bring advanced nanomaterials to the market in the form of new products and applications.

The BMS experience over the past decade with CNTs is a clear example. Nobody disputes the theoretical properties of advanced nanomaterials such as CNTs and graphene. These are well known. As Andrew Geim recently put it: Graphene is dead. Long live graphene! Hundreds of peer-reviewed scientific papers have been published on the properties of graphene and other nanomaterials. The major issue associated with these materials is not theory and properties, but practice and application. How do we turn their fantastic properties into useful and, in some cases, game-changing products?

It is clear from the experience of BMS and others that traditional approaches to commercializing these materials are not effective. The emergence of MPM is due to the shortcomings of these traditional approaches. We know that growth is a function of learning. After all, the cave man had access to all of the materials we have today. What the cave man did not have was the propensity for learning that comes from having experienced failure and success. MPM embraces the learning curves associated with bringing advanced nanomaterials to the market through the development of new processes, standards, tools, and technologies.

There is no reason a priori to expect the earlier-described TSPPs associated with the successful commercialization of non-nanomaterials to be the same for nanomaterials. It is natural to want to apply the same tried-and-true TSPPs to commercialize advanced nanomaterials. At the heart of MPM is the development of new TSPPs necessary for the proper characterization and functionalization of advanced 2D and 3D nanomaterials, together with its effect on the target matrix and down-stream processing.

"Characterization" of nanomaterials is critical. Characterization involves the use of sophisticated metrology tools and information technology that peer down into the nano world and generate data that help us identify the type of nanomaterial being developed for commercialization. Manufacturers today might believe they are working with graphene because their supplier told them it was graphene, when in truth, characterization identifies the material as akin to "soot." And there is a world of difference between graphene and soot. Knowing the kind of material one is using is paramount to the commercialization process. The way to know what type of material is being used is via characterization analysis. Characterization analysis enables material comparison and is a key component – and the foundation – of the MPM paradigm.

A great deal of work is being done today by researchers at the National Physical Laboratory (NPL) in the United Kingdom and elsewhere that is pushing the envelope of characterization analysis. NPL and others are pioneering new techniques that allow for more accurate assessment of nanomaterials, and even tools to enable real-time characterization of graphene. New types of metrology tools are being developed to foster characterization analysis of newly discovered 2D nanomaterials.

Researchers at Lancaster University (LU) note that scanning probe microscopy (SPM) represents a powerful tool which, in the past three decades, has allowed researchers to investigate material surfaces in unprecedented ways at the nanoscale level. However, SPM has shown very little power of penetration, whereas several nanotechnology applications would require it. The LU researchers are using other tools, such as ultrasonic force microscopy (UFM), in work with graphene and other 2D materials, including MoS_2. UFM is a variation of the atomic force microscope (AFM) that overcomes the limitations of SPM in characterizing advanced nanomaterials such as graphene and other 2D materials.

These new tools and techniques in development will give manufacturers the important data necessary to ensure that the correct material is being used in the manufacturing process. They also promise to foster quality control in a manner that has not existed previously. As producers in any industry know, quality control is paramount to successful commercialization. Additionally, the creation of sophisticated models to assist in the development, design, and integration of these materials into devices and products relies heavily on the completeness and reliability of property data for these nanomaterials.

Characterization work also facilitates the development of standards that are critical to the evolution of advanced nanomaterials. The term graphene today covers a family of different materials, including several-layer flakes, powders, liquid dispersions, and graphene oxide. Importantly, the corresponding properties and potential applications will vary depending on the type of material used.

The other critical part of MPM is dispersion. The ability to consistently and uniformly disperse graphene in another material is important to realizing the outstanding properties of the material. Functionalizing graphene properly can enhance the strength, stiffness, and conductivity of the resulting composites, depending on the requirements and applications being targeted.

1.4 Addressing the Environment, Health, and Safety

Another important component of the emerging MPM paradigm relates to the environmental, health, and safety (EH&S) procedures and protocols for advanced nanomaterials. There have been a number of "scare stories" in the media about the potential toxicity of various nanomaterials. Most of these fail to consider the final product form that nanomaterials actually take when introduced to the market, as well as the potential, or lack thereof, of their release into the environment as nano-sized particles.

Without a clear understanding of the full manufacturing cycle, product form, and disposal considerations, the limited information generated by current studies is of little relevance. Additionally, lacking test standards and precise definitions, it is impossible to conduct credible, repeatable, and scientifically valid studies. All of the characterization work that is going on behind the scenes with graphene and other 2D materials today is important to future EH&S studies.

It is incumbent upon all in the nanomaterials community to collaborate on EH&S-related issues. The new characterization tools and techniques that have been developed and are being developed will help facilitate toxicity studies. There are groups of researchers today, such as the Arkansas Research Alliance, that are intent on doing credible nontoxicity research on graphene and other nanomaterials that can be of benefit to all who wish to promote the responsible development of such materials.

One way to minimize the EH&S effect and aid commercialization is to add the nanomaterials to a carrier in the form of a loaded masterbatch, which is then let down (diluted) by a processor with the raw, untreated carrier material. This offers controllability; and once in a masterbatch, it can be handled without the need for expensive nano-handling environments.

1.5 The Nanomaterials Road Ahead

We are still at an early stage with the new MPM paradigm. The promise of nanomaterials such as graphene and CNTs is great, but so, too, are the challenges associated with successful commercialization. Several of the key challenges associated with commercializing nanomaterials-enabled products are being addressed through the development of the MPM paradigm. Again, considerable progress has been made, but there is much more work to be done in terms of testing and data analysis.

Companies seeking to work with graphene and other nanomaterials need to know the type of materials they are using. Characterization analysis provides this information and also helps to facilitate standards that are necessary for industry maturity and EH&S-related research. Additionally, companies need ways of reliably producing materials to achieve their desired properties. Functionalization assists greatly in this area, for without it, the inert carbon-based material will not want to disperse readily into a target medium. With respect to functionalization, it is also early days, but we see a great deal of potential as functionalization becomes commonplace among those commercializing advanced nanomaterials. It is clear from the lack of progress with CNTs thus far that there is a need for a paradigm such as MPM if we are going to realize the promise and potential of graphene and other nanomaterials.

The excitement over these newly discovered nanomaterials is warranted, but again, those seeking to invest and innovate in this promising area need be mindful of the challenges associated with commercializing these materials. Key to progress on the commercialization front is close collaboration among suppliers and producers and a good deal of patience among all participants involved: the history of materials tells us that it can take years, and sometimes decades, before a new "wonder material" fulfills its promise and potential.

Consider the evolution of materials such as aluminum and advanced ceramics. Aluminum was discovered in a lab in the 1820s. Like CNTs and graphene, the material was hailed as a wonder substance, with qualities never seen before in a metal. However, it proved expensive to make, and it was not until many decades later that it took off in the marketplace, when a new process using electricity was invented.

Similarly, many of us remember the excitement surrounding advanced ceramics in the early 1980s, and the fever that developed with the discovery of high-temperature ceramic superconductors. The promise of ceramic engines, loss-free electrical transmission lines, and many other products that these material advances were expected to enable has remained unfulfilled. That said, the impact that these materials have had on our lives is nearly impossible to list – ranging from the mundane to the exotic and impacting transportation, communications, electronics, consumer goods, medical devices, and energy in ways that may be hidden but are enabling nonetheless.

The road ahead for the development of applications and products using 2D and 3D nanomaterials is filled with tremendous opportunities and key challenges. There is also always a great deal of hype surrounding the discovery of new materials, and experience teaches that hype often turns to disappointment before a wonder material's potential is eventually fulfilled. In the main, those who earlier tried CNTs and failed remain willing to experiment with the likes of graphene and other nano materials as the desire to get a competitive advantage remains a key economic driver in a very competitive world.

The emerging MPM paradigm discussed in this paper seeks to foster the acceleration of the commercialization process of advanced nanomaterials and promote their responsible development. The TSPPs are designed to avoid corporations from being tempted to reach for instant volume in a desire for market dominance, growth, and profit. The investor community needs to be wary of those who claim volumes that are in the many tons, or hundreds of tons, without proving scale-up as well as process controls to ensure consistent quality production. For those who seek instant "glory," the bear trap of failure through nonrepeatability looms large.

Despite the great deal of work ahead to realize the potential of these exciting materials, and despite some of the setbacks encountered over the past decade, we are encouraged by the progress we are making to bring these next-generation "wonder materials" to the market.

1.6 Can Graphene Survive the "Disillusionment" Downturn?

Even if you are not familiar with the life cycle of emerging technologies, you have certainly heard about technologies that generated lots of interest at an early stage but a few years later are gone, having never really entered the marketplace. Many of these technologies showed outstanding results in the lab but were unsuccessful in moving out of the lab into the real world. Most tech companies must pass through the ups and downs of their industry's life cycle, but how they understand and react to these cycles can make a big difference.

1.6.1 Gartner's Hype Cycle

The Hype Cycle is a branded graphical tool developed by the research and advisory firm Gartner (www.gartner.com) for analyzing the maturity and adoption of emerging technologies.

Technology X (a shiny, life changing, and innovative tech) is introduced as the next big thing (Technology Trigger) and everyone is talking about how it will change our life (Peak of Inflated Expectations)! Then, as reality sets in, people realize that everything has not magically changed and disappointment sets in (Trough of Disillusionment). The shiny, new technology starts to look dull. As time goes by, smart people look at the real opportunities for the shiny new technology (Slope of Enlightenment) and learn how to build solid businesses with the not-so-shiny-and-new thing (Plateau of Productivity). This is how technology X goes from the lab to the real market (Figure 1.1).

The period of time from discovery to maturity is variable and depends on the type of technology; for instance, it takes around 25–30 years for a new advanced material to move through the cycle. The best recent example is CNT, graphene's sister material, discovered in 1991. Today, after 23 years, the CNT industry is slowly moving up the "Slope of Enlightenment." Graphene will pass through a very similar cycle, although the cycle time may be slightly faster since many graphene players have learned from CNTs' hurdles.

Graphene was discovered in 2004, and the first generation of graphene producers, such as XG Science, Angstron Materials, and Vorbeck Materials, had launched and introduced their first generation of products by 2008. During early 2010, large corporations such as BASF (early adopters) showed interest in graphene and began to test first-generation products. Results were often disappointing due to problems with graphene dispersion, lack of batch-to-batch consistency, and the lack of clear graphene standards.

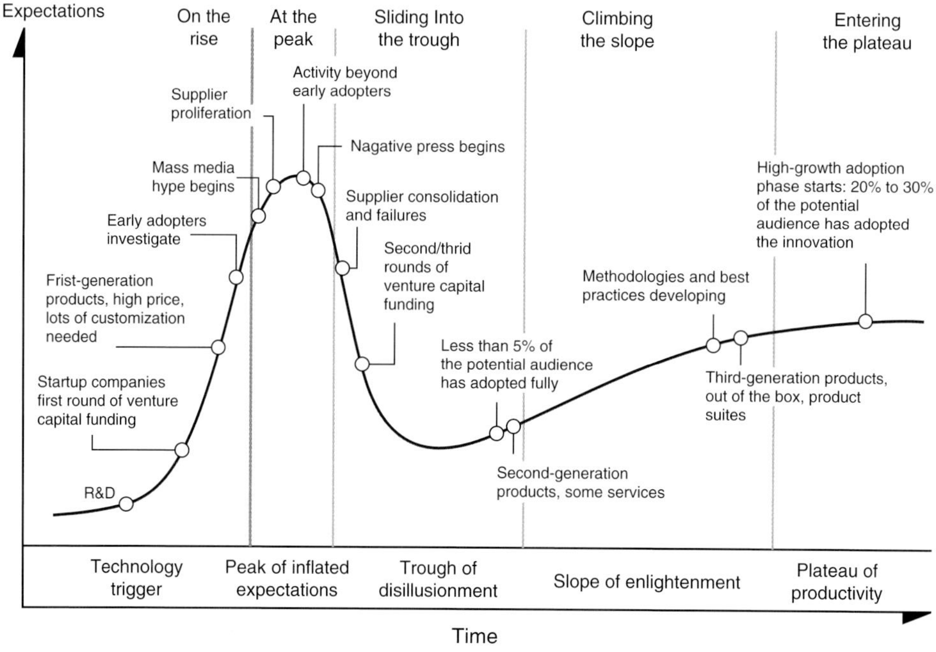

Figure 1.1 Gartner's hype cycle.

In September 2010, Konstantin Novoselov and Andre Geim were delighted to receive the Nobel Prize in Physics for the discovery of graphene. This award resulted in broad media coverage, building up to mass media hype by early 2011. Media hype continues today as governments launch and build support for large science to industry programs. After few years of excitement and buildup, graphene is now "At the Peak" of Gartner's Hype Cycle and sliding into the "Trough of Disillusionment" is happening.

1.6.2 Surviving the Trough of Disillusionment

As scary as the "Trough of Disillusionment" appears, there are a few key strategies that graphene companies can employ to safely move through this stage:

1) **Maintain low overheads**. Growing too fast and burning cash at the "Peak of Inflated Expectations" stage has killed a lot of businesses. Access to capital becomes much harder as an industry moves into the "Trough of Disillusionment."
2) **Concentrate**. Graphene companies need to focus on one target market in which their products provide the maximum value to their customers. Trying to chase all opportunities, across multiple industries, increases the burn and reduces the chances of success dramatically.

3) **Be revenue-driven, rather than value-driven**. Many potential markets (e.g., bio/health, aerospace) have a huge upside in terms of value, but time-to-market is long and regulatory hurdles exhausting; lots of cash will be burned before significant revenues are made. Unless graphene companies have a strong partner with deep pockets, short-term revenue opportunities must trump long-term, value-driven markets.

Investors on the other hand, need to understand that the development of an advanced material business is a long process and depends heavily on management strategy. Companies that can ignore the hype, and grow and generate revenues during the initial industry phases, will have the opportunity to create lasting, valuable businesses. Reviewing the life cycles of other high-tech markets, such as those of solar cells and plastics, may help provide investors with key insights.

1.6.3 Graphene and Batteries

Researchers note that graphene can improve battery attributes like energy density and form in various ways. Conventional battery electrode materials, as well as prospective ones, can be significantly improved when enhanced with graphene. Graphene's unique traits, such as mechanical strength, electrical conductivity, large surface area, and lightness of weight can make batteries lighter and more durable and suitable for high-capacity energy storage. Additionally, graphene can shorten charging times – a highly desirable feature for electric vehicles (EVs) and consumer electronics products. A battery's lifetime is negatively linked to the amount of carbon that is coated on the material or added to electrodes to achieve conductivity (Figure 1.2). Graphene adds conductivity without requiring the amounts of carbon that are used in conventional batteries.[1)]

Graphene, and in particular graphene oxide, has shown to be a valuable material for overcoming the hardest challenges presented in lithium–sulfur batteries. In summer 2014, researchers from Samsung's Advanced Institute of Technology (SAIT) announced a novel way to extend the life of a lithium-ion battery (LIB) using a combination of silicon and graphene. SAIT fabricated anode material by growing graphene on the surface of silicon without forming silicon carbide. The new material has four times the capacity of commercial graphite. Researchers at SAIT note that the approach has the potential to increase the volumetric energy density of LIBs by 1.8×. Key to the commercialization of this advanced graphene-enabled battery technology is the ability to manufacture carbide-free graphene in mass quantities. The biggest obstacle to realizing the full application of graphene technology today is the relatively high cost and low reliability for large-scale production and manufacturing. There is a great deal of work being done in South Korea by Samsung as well as by other researchers inside and outside large established corporations around the world to address this critical issue.

1) For more, see Roni Peleg and Ron Mertens, *Graphene Batteries Market Report*, 2015.

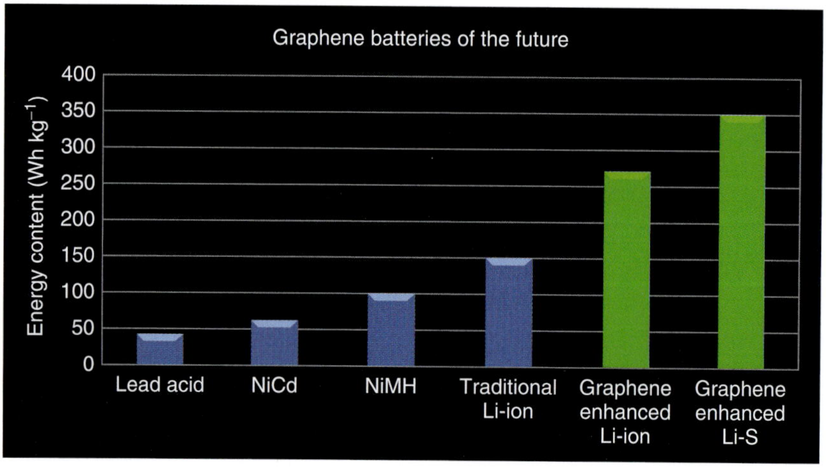

Figure 1.2 2-D nanomaterials enable more powerful batteries. (Peleg and Mertens, *The Graphene Batteries Market Report*, 2015.)

LIB is the most important type of rechargeable batteries. In such types of batteries, lithium moves back and forth between two electrodes, called cathode and anode, for charging and discharging. LIBs are common in many consumer electronics and electric cars due to their relatively high energy density (the amount of energy stored in a unit of battery), low hysteresis (after charging and discharging, there is little loss of energy capacity), and a very slow loss of energy when not in used. LIBs consist of a lithium compound as cathode, spherical graphite as anode, and lithium salt as an electrolyte to allow lithium ion movement between the cathode and anode. Increasing the capacity of LIB is dependent upon better materials for cathode and anode. It should be noted that the combination of cathode, anode, and electrolyte is one cell; several connected cells are called a *module and multiple modules* go together to make up a battery.

Recently, news regarding the proposed Tesla battery Gigafactory has had an impact on the industries involved in the LIB supply chain, notably on natural flake graphite junior miners. A large component of today's LIBs is graphite and, for the proposed Tesla factory only, more than 300k metric tons/year graphite would be needed. The news of the proposed Gigafactory has resulted in a boost in the graphite market, but graphite-based anodes are not at all adequate for the battery performance that is required for EVs by 2030. By that date, most hybrid electric cars will have been converted to full electric cars running completely on battery power and without any fossil fuel consumption. The replacement material has to radically improve the performance of existing batteries to provide longer run times (a larger storage of energy), faster charge times, all with the smallest possible weight and at the lowest possible added cost. Furthermore, the new batteries need to be long lasting (over 1000 cycles) and thermally stable (should not be over-heated during charging). Graphene is a leading candidate for the replacement material.

There are many studies and technical papers showing how graphene can improve batteries. Its outstanding electrical and thermal conductivity enhances the activity of cathodes and prevents over-heating of the batteries. Recent findings by researchers from Lawrence Berkley lab introduced lithium–sulphur graphene compounds that generated twice the energy of current batteries and were stable over 1500 cycles. Such batteries could enable EVs with an efficiency of more than 500 miles on a single charge, which is what future electric cars need. Newer technologies such as Li–air batteries or supercapacitors could replace LIBs as well.

The future of the energy industry is largely dependent upon improved batteries. Such batteries will change our life drastically. In a matter of few years, gas stations will be replaced by electric car charging stations and typical auto mechanics will require new certification to repair electric cars. Further investigation on graphene-enhanced batteries is absolutely crucial as graphene–silicon compounds have proved to be a potential replacement for spherical graphite as anode in LIBs and graphene oxide–sulphur compounds as cathode in lithium sulphur batteries.

1.6.4 Heat Management with Graphene

Miniaturization of electronic systems and circuits is heavily restricted with heat dissipation challenges. Heat buildup reduces the efficiency of the electric motors, performance of CPUs, and lifetime of consumer products and batteries. Heat dissipation becomes even more challenging when flexibility and bendability of the final product is important. Metals are not a suitable candidate anymore and plastics are rapidly replacing them as they are cheaper and easier to shape and are weightless. However, plastics severely suffer from lack of thermal conductivity. The possibility of enhancing thermal conductivity of plastics (preferably by keeping them electrically insulating) is game changing. Graphene is proved to have the highest thermal conductivity among all materials. Small loading of well-dispersed graphene into plastics can enhance their ability to dissipate the heat tremendously. If such a loading level is lower than the percolation threshold, plastics stay insulating. Percolation in plastics starts by 0.2–0.5 wt% of graphene addition. Lower concentration of graphene is likely to be ineffective to change electrical conductivity. Table 1.1 represents the impact of graphene addition upon the thermal conductivity of thermoplastics and thermosets.

Having said this, selection of the optimized graphene loading is crucial. Concentration of graphene has to be finely tuned to an optimal value in order to achieve the best results. For instance, in case of poly(lactic acid) (PLA), the optimal concentration of graphene was found to be 0.075% (Figure 1.3).

Table 1.1 NanoXplore graphene improves polymer thermal conductivity and effusivity.

Material	Thermal conductivity at 21–25 °C ($W\,m^{-1}\,K^{-1}$)	Improved thermal conductivity	Thermal effusivity ($Ws^{0.5}\,m^{-2}\,K^{-1}$)	Improved thermal effusivity
PLA[a)]	0.36	245%	714	112%
PLA + 0.075 wt% graphene	1.23		1517	
PE[b)]	0.74	44%	888	55%
PE + 0.1 wt% graphene	1.06		1377	
ABS[c)]	0.29	339%	643	142%
ABS + 0.05 wt% graphene	1.28		1555	
Silicone rubber	0.23	446%	572	166%
Silicone rubber + 0.2 wt% graphene	1.24		1522	
2-Part epoxy potting compound	0.38	45%	771	17%
2-Part epoxy potting compound + 0.075 wt% graphene	0.55		905	
Silicone heat transfer compound	0.66	54%	1190	15%
Silicone heat transfer compound + 0.1 wt% graphene	1.02		1367	
Polyurethane	0.21	80%	550	33%
Polyurethane + 0.13 wt% graphene	0.37		730	

a) PLA stands for Poly(lactic acid).
b) PE stands for Polyethylene.
c) ABS stands for Acrylonitrile butadiene styrene.

1.6.5
How Graphene Could Revolutionize 3D Printing

Last year at the International Manufacturing Technology Show (IMTS) in Chicago, one of the largest industrial trade shows in the world with more than 100 000 visitors, 1900 exhibitors gathered in Chicago to showcase recent developments in machines, tools, and manufacturing systems. Arizona-based automobile manufacturer "Local Motors" stole the show by printing and assembling an entire automobile, called the *Strati,* from scratch and live in front of spectators. On the other side of the world, a Chinese company "WinSun Decoration Design Engineering" recently constructed a set of 10 single story, 3D-printed homes produced in under 24 h. These homes, based upon cement-based prefabricated panels printed on a custom-built 10 x 6.6 m 3D printer, were assembled on site

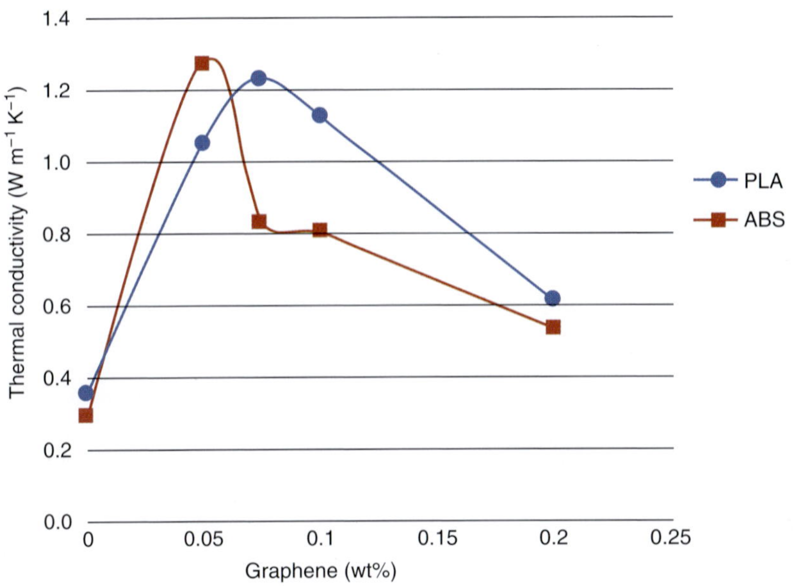

Figure 1.3 Thermal conductivity of thermoplastic/graphene in different loading of graphene.

and cost just $5000 to build. These are just two concrete examples of the recent rapid progress in the 3D printing industry.

Three-dimensional printing, also known as *additive manufacturing*, is the process of using successive layers of printed material to form solid 3D objects of virtually any shape from a digital model. Specially formulated materials, such as plastics or powdered metals, are used to build up successive layers to create components with complex shapes. The final characteristics of the 3D printed piece can be modified depending upon the additives used in the printed material and the printing techniques. Plastics are the most versatile printing material, but they are not mechanically strong and lack thermal and electrical conductivity, which is a requisite for many applications in the electronic and aerospace industries. In 2014, sales of industrial-grade 3D printers in the United States had risen to a level equivalent to one-third the volume of industrial automation and robotic sales. Sales of such printers are expected to continue to increase measurably in the years ahead due to advances in 3D printing software and the development of new printable materials such as graphene, CNTs, and other advanced nanomaterials.

Graphene, for example, has recently been explored for the printing of 3D structures of various dimensions having controlled properties. Example applications include printed electronics, biosensors, strain sensors, battery electrodes and separators, or filtration wherein the electrical, physical, chemical, or mechanical properties of the structures are controlled to provide targeted functionality by design. Utilizing processes such as inkjet or nanoimprint lithography, structures have been realized for printed electronics and sensors. More recently, a 3D printing strategy has been demonstrated for the fabrication of 3D graphene aerogels with designed macroscopic architectures, enabling a method to further control the mechanical and surface area properties of complex macroscale structures.

One of the main technical challenges to adding graphene to plastics is graphene's limited dispersibility, or its relatively poor ability to mix with other materials. Dispersibility of graphene in polymers requires special attention to the graphene's edge structure. Depending on the type of plastic, the graphene edge must be functionalized by adding specific molecules, which enables effective bonding between graphene and plastic. The severity of mixing problems of CNTs, a sister material to graphene, eventually nearly killed its market and forced many CNT producers to abandon their businesses. Graphene performs much better than CNTs in terms of dispersion; yet, ironically, the best quality graphene, with high purity and excellent crystalline structure (low defects), has the poorest mixing performance. Oxidized graphene, with lots of oxygen molecules available for bonding, provides better mixing with plastics, but large-scale production of graphene oxide is very challenging. Production of graphene oxide by the Hummer's method, the industry standard, is expensive and complex as it needs to guard against explosions and manage the use of large volumes of acids and harsh materials. The future of graphene additives for 3D printing requires either new approaches to managing the graphene edge structure or a new approach to manufacturing graphene oxide.

The value proposition for additive manufacturing is compelling today and likely to get more attractive in the months and years ahead. A significant part of the advantage of 3D printing is related to being able to manufacture single parts in one run, which can dramatically cut the cost of production. GE's Aviation division is 3D printing a fuel nozzle that previously involved assembly from 20 separately cast parts. GE found that the new additive process reduced the cost of manufacturing by a whopping 75%. Ducati is using 3D printing to substantially reduce product development time and to increase efficiency.

Additive manufacturing methods can be used to combine parts and foster greater detailing. They can also use multiple printer jets to lay down various materials simultaneously. There is also the ability to functionalize materials and create an array of innovative, customized products with unique features. Researchers today are using 3D printing to create customized body parts that can be used to replace human organs and limbs. With additive manufacturing, science fiction is becoming science fact. One envisions companies all over the planet having to redesign and reengineer their manufacturing processes around additive manufacturing and the processes associated with it. 3D printing with advanced nanomaterials such as graphene hugely enables technological innovation inspired by nature (i.e., biomimicry). The convergence of atoms (nanomaterials and hardware) and bits (3D printing software) represents a vast frontier of transformative deep science-enabled innovation attractive to companies and venture investors alike. When it comes to 3D printing, new breakthroughs and new achievements are being realized almost on a daily basis. Graphene is a great additive material and has significant potential to revolutionizing the range of possible products that can be manufactured with 3D printing. However, adding a complicated

Table 1.2 NanoXplore graphene improves polymer mechanical properties.

Material	Ultimate tensile strength (MPa)	Tensile strain at break (%)
Base rubber compound	11.70	540.56
Base rubber compound + 0.1 wt% graphene	12.99	663.18
Acrylated monomers	19.5	0.71
Acrylated monomers + 0.5 wt% graphene	30.6	1.23

Additive manufacturing is a game changer for industry.

material such as graphene will require significant effort and patience to perfect the technology and achieve its promise.

An example of such improvements are enhancing the mechanical properties of rubber-based filaments and UV curable filaments. Table 1.2 represents some of the improvements.

As we can see, there are many exciting opportunities that have arisen along with the discovery of graphene and other 2D nanomaterials in the past decade. This book provides an overview of some of the important ongoing research with graphene and also highlights some of the commercial trends and related issues associated with financing companies innovating with 2D materials. The past decade has been one of intense research in 2D nanomaterials. As we have seen with CNTs and other advanced materials, the commercialization cycle extends out over a decade. Experience informs us that where there is great opportunity for commercialization with 2D nanomaterials, there are also challenges and risks associated with creating sustainable business models and successful companies. Based on their remarkable properties and ongoing research and development trends highlighted in this book, we are optimistic about the commercialization prospects of graphene and other 2D nanomaterials-enabled products in the years ahead.

2
Graphene Synthesis

Siegfried Eigler

2.1
Introduction

Graphene is exactly one carbon layer of graphite. The carbon atoms are arranged in a honeycomb lattice. However, the access to graphene is one hurdle for the development of applications based on graphene. In general, there are two approaches to graphene. One starts from graphite (top-down approach), the other one from small molecules that are used to build up graphene (bottom-up approach). Not only is the isolation of one layer of graphite a challenge but so also is the stabilization of the delaminated layers of graphene. If layers of graphene are not stabilized they tend to aggregate, forming a sort of graphite. While there is a stacking order in graphite, restacked graphene exhibits no order of layers. Instead, the flakes of restacked graphene are randomly stacked and it seems likely that porous structures will be formed, very similar to crumpled paper piles.

The delamination of graphite to graphene can be performed mechanically using adhesive tape, and mechanically in solvents by sonication, shear mixing, or ball milling. In addition, the delamination of single layers of graphite can be facilitated using chemical methods that involve chemical functionalization and defunctionalization after processing.

Another approach to graphene focuses on the synthesis of graphene on solids using a carbon source. The method of choice is most often chemical vapor deposition (CVD) using a metal surface and small molecules, such as methane or acetylene. The challenges in the CVD approach are that the substrate should meet some requirements: the processing parameters such as temperature, gas mixture, and partial pressure, as well as the subsequent processing of grown graphene. Until now, the synthesis of graphene has been demonstrated by both non-wet-chemical methods and chemical methods. However, processing and integrating graphene in either electronic devices or composite materials to obtain high performance materials by using the unique properties of graphene, such as the high mobility of charge carriers, mechanical strength, conductivity, and transparency combined with flexibility is another challenge that must be overcome in the future.

Graphene Technology: From Laboratory to Fabrication,
First Edition. Edited by Soroush Nazarpour and Stephen R. Waite.

The challenge that all methods for the synthesis of graphene have in common is that the quality of graphene differs, which is due to the polydispersity of graphene. Thus, when talking about graphene, not only the synthetic method but also the resulting graphene in terms of the size of graphene, which can vary between nanometers, micrometers, and continuous films must be taken into account. Furthermore, the density and type of defects can vary strongly depending on the type of processing. Although graphene is a single layer, often more than one layer is obtained. In addition, graphene can be chemically modified or functionalized by covalent or non-covalent methods. The degree of functionalization is another polydispersity that should be taken into account when discussing graphene.

2.2 Definitions

2.2.1 Nomenclature and Structure

Graphene is exactly one layer of graphite [1]. The C-atoms of graphene are arranged in a honeycomb lattice and the C−C bond distance in graphene is 0.142 nm. In addition, graphene is an infinite plane. However, real graphene exhibits edges that can either be zigzag or arm-chair (Figure 2.1) [2].

Bi-layer graphene consists of two layers of graphene stacked on each other. Generally, it is not distinguished how they are stacked, but in principle they can be stacked as AA or AB. In AA stacked bi-layer graphene, the layers are stacked in a way that carbon atoms are on top of each other. In AB stacked bi-layer graphene, carbon atoms of the second layer are on top of the cavity of a six-membered ring of the first layer. Besides this ideal stacking, the layers can also be twisted to form a moiré pattern that is dependent on the twisting angle (Figure 2.1) [4]. Tri-layer graphene can be formed along these lines and up to 10 layers may be called few-layer graphene. However, as it is often difficult to distinguish between, say, 8 or 10 layers, the term multi-layered graphene can also be used. The properties of more than 10 layers of graphene are similar to those of graphite and therefore, the term graphite is used. However, the layer sequence may also be random, and is termed turbostratic graphite. Restacked graphite or also porous graphite is yielded when layers of graphene or few-layer graphene restack after delamination or partial delamination (Figure 2.1).

2.2.2 Polydispersity of Graphene

In contrast to the fullerene C_{60}, which is a molecule and a monodisperse material, graphene is a polydisperse material. Even if C_{60} is contaminated by C_{70} after synthesis, C_{60} can easily be purified by chromatography. Such simple procedures are not available for graphene and graphene-based systems. The following

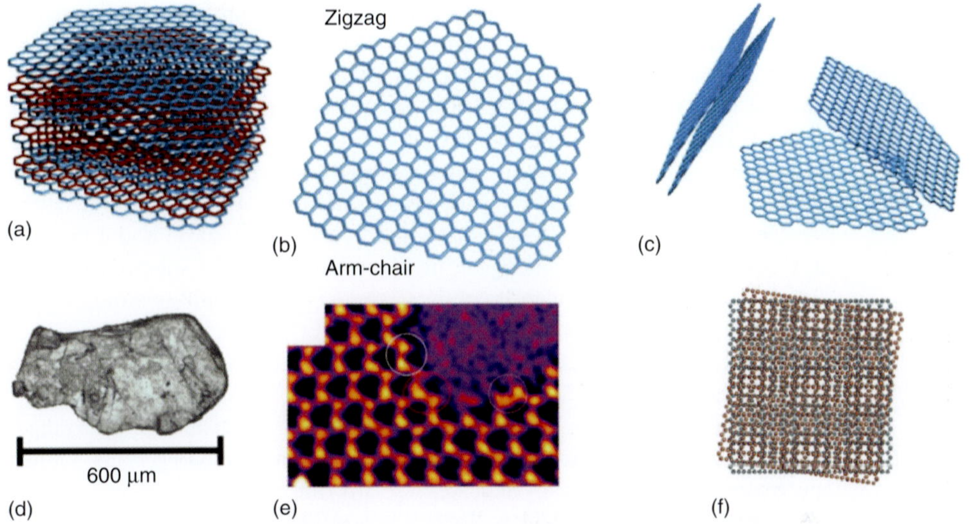

Figure 2.1 (a) Chemical structure motive of AB stacked graphite. (b) Graphene with zigzag and arm-chair edges. (c) Restacked layers of graphene and few-layer graphene. (d) A flake of natural graphite. (From Ref. [2] with permission from Wiley-VCH Verlag GmbH & Co, Copyright 2014.) (e) High resolution transmission emission spectroscope (HR-TEM) image of one layer of graphene. (Reprinted from Ref. [3] with permission from Macmillan Publishers Ltd: Nature Communications, Copyright 2014.) (f) Selected moiré pattern of twisted bi-layer graphene. (Reprinted from Ref. [4] with permission from Macmillan Publishers Ltd: Nature, Copyright 2011.)

equation expresses some of the possible aspects of polydispersity of graphene and graphene-based systems [2, 5]:

$$S/^{s,d}G_n - (R)_f/A_f \qquad (2.1)$$

S is the substrate, s the size of graphene, d the structural defect density of graphene within the carbon framework, G the graphene, n the number of layers of graphene, R the addend, f the degree of functionalization, A the non-covalently bound molecules, and for excluded S, reactions are applied in dispersion.

First of all, it is important to distinguish between graphene, few-layer graphene, and more layers. Therefore, graphene (G) is used in Eq. (2.1) with an index n indicating the number of layers. Graphene (G_1) is used for graphene, and $G_{\text{few-layer}}$ or G_{3-10} indicates few-layer graphene. An infinite number of layers can be indicated either by G_∞ or by G_n signifying graphite. Moreover, graphene may be stabilized or deposited on a solid support such as a surface. An often used surface is SiO_2 grown on a Si wafer. Other surfaces are BN (boron nitride) or that of a polymer. The deposition of graphene on SiO_2 can thus be indicated by SiO_2/G_1. Alternatively, graphene may be deposited on a surface of water, for example by etching copper away after growth of graphene on copper, which is indicated by H_2O/G_1. Another type of polydispersity is the size of flakes, which can be on the nanometer-scale, micrometer-scale, or even centimeter-scale. Exemplarily, graphene grown on a copper foil of a size of 1 cm × 1 cm can be indicated as $Cu/^{1\,cm}G_1$. Flakes of graphene, delaminated from graphite with a size of the flakes of about 5 μm and placed on a SiO_2 wafer, are indicated as $SiO_2/^{5\,\mu m}G_1$. Another issue is about defects within the carbon framework of C-atoms. This means that the honeycomb lattice is disturbed. The density of these defects can be determined by the analysis of results obtained by Raman spectrascopy. Raman spectroscopy is very sensitive toward defects and the density of defects can be determined in a range of about 0.001% and 1–3%. It should be noted that not all types of defects in graphene can be identified by Raman spectroscopy and therefore, the determined density of defects must be seen only as a rough value [6]. The type of defects and the shape and size of defects are also difficult to determine and these types of polydispersity are currently under investigation in basic research.

Structural defects can be due to missing atoms or mismatching arrangements of C-atoms, like five-membered rings or seven-membered rings. Raman spectroscopy however, also indicates sp^3-hybridized carbon or clusters of sp^3-hybridized carbon as defects. Thus, it is possible to determine the degree of functionalization by Raman spectroscopy, especially for graphene with an intact carbon framework [7]. The addition of functional groups to the lattice of graphene will lead to functionalized graphene, which reflects another type of polydispersity of graphene. The regiochemistry of addends on graphene is, until now, almost unexplored and therefore, the regiochemistry is normally not indicated. However, the *trans*-1,2-addition motive of functional groups is thermodynamically most favored. However, reactions with graphene placed on surfaces can only proceed from the upper side. In the systematic of Eq. (2.1), the degree of functionalization based on the available C-atoms is indicated. For every 20th C-atom being functionalized, for example, by OH, the degree of functionalization is 5% and therefore, is indicated by $G_1\text{-}(OH)_{5\%}$. Furthermore, few layers of graphene can be exfoliated from graphite in stabilizing solvents such as *N*-methyl-2-pyrrolidone (NMP), which results in adsorption of NMP on the surface of graphene. Such a type of graphene is indicated by G_1/NMP and if the amount of NMP is determined, for example, by elemental combustion analysis (determination of N-content), the mass % content can be indicated by an index ($G_1/NMP_{30\%}$).

It is of upmost importance to characterize synthesized materials in detail and the results must be indicated in a systematic way. Otherwise, it is difficult for the readers to realize what type of material was exactly produced. Therefore, the expression given in Eq. (2.1) makes it possible to understand the relationship between the structure of graphene and the determined properties.

2.3 Characterization of Graphene by Raman Spectroscopy

Raman spectroscopic characterization of graphene is the most reliable method to determine the quality of graphene by identifying the density of defects within

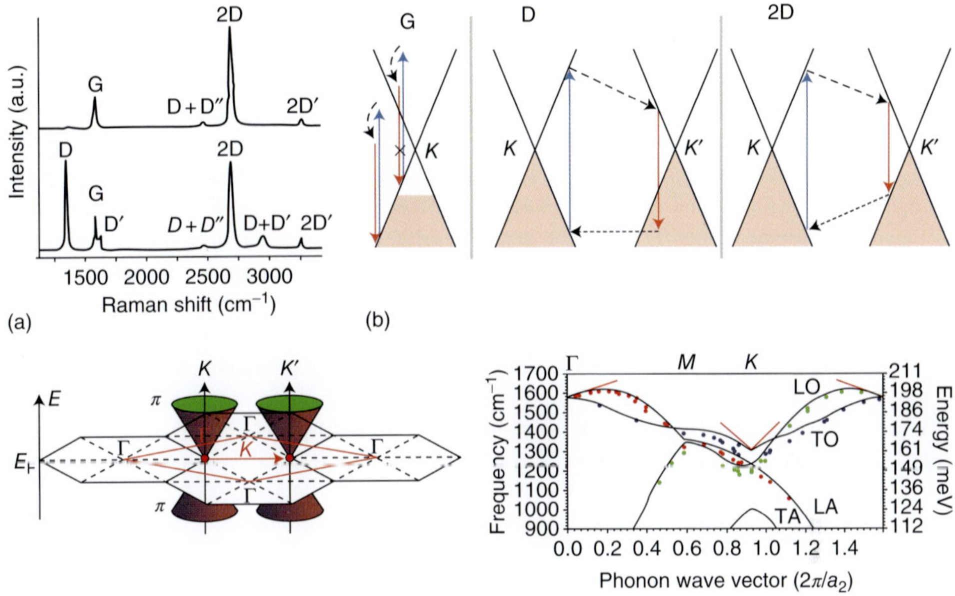

Figure 2.2 (a) Raman spectra of graphene (top) and graphene with defects (bottom) with the most relevant peaks labeled with D, G, and 2D. (b) Illustrations of selected excitation and emission processes responsible for the G, D, and 2D peak. (Adapted from Ref. [6] with permission from Macmillan Publishers Ltd: Nature Nanotechnology, Copyright 2013.)

the C-framework. Raman spectroscopy of graphene can be performed by using several laser wavelengths ranging from the blue to red region of the light spectrum.

However, a green laser turned out to be the most popular one. Either 514 or 532 nm of excitation is used. The Raman spectrum of graphene displays three major peaks that can be used to evaluate the quality of graphene. These peaks are the G peak, D peak, and the 2D peak (Figure 2.2). The G peak at about 1580 cm^{-1} is an allowed Γ-point phonon emission and its intensity scales with the number of C-atoms probed by the laser spot. The D peak at about 1345 cm^{-1} is a result of phonon emission that is forbidden by selection rules and is only possible due to the activation by defects. These defects can be those within the carbon framework, such as missing atoms or substituted atoms. Alternatively, the transitions of sp^2-carbon to sp^3-carbon also display a defect that can be probed by the D peak [7]. Furthermore, the D-peak position is dispersive, which means that its position shifts with the excitation wavelength of the laser. The second order of the D peak, the so-called 2D peak, is the consequence of an allowed phonon process and the 2D peak is the most intense peak in graphene. The shape of the 2D peak can be fitted with a single Lorentz function and the full-width at half-maximum (FWHM, Γ) is about 30 cm^{-1}. For bi-layer graphene, the $\Gamma_{2D} > 40\ cm^{-1}$, which allows distinguishing between graphene and few-layer graphene [8]. As mentioned above, the D peak evolves with the introduction of defects. While the intensity of the G peak remains almost constant, the intensity of the D peak rises, which results in an increased I_D/I_G ratio. The maximum ratio possible is dependent on the excitation laser wavelength. For an excitation at 532 nm the I_D/I_G ratio increases to a maximum of about four. With further increase in the density of defects the I_D/I_G ratio declines until the I_D/I_G ratio is of about one. This declining of the I_D/I_G ratio is accompanied with line broadening. Thus, the I_D/I_G ratio follows a relation and therefore, an I_D/I_G ratio of two can be related to two different densities of defects. By taking the Γ of peaks into account it is possible to distinguish between both. By introducing defects, Γ_{2D} increases from 30 cm^{-1}, for a high quality of graphene, to about 50 cm^{-1}, which relates already to a distance of defects of about 3 nm. The Γ_{2D} can further increase to values up to 350 cm^{-1}, which means that the 2D peak is only barely detectable. This also means that the structure of graphene is destroyed and several percentage values of defects are introduced within the carbon framework (Figure 2.3) [9, 10].

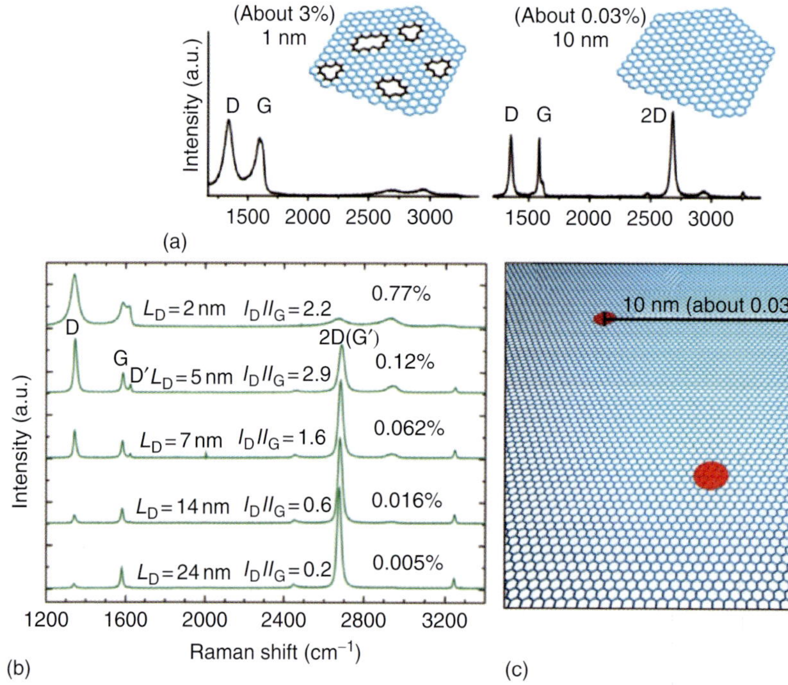

Figure 2.3 (a) Raman spectra of graphene with different densities of defects. Left: defect density about 1–3% (no sharp 2D peak); right: density of defects of 0.03%. (b) Reference Raman spectra of graphene that relate to densities of defects between 0.005% and 0.77%. (Reproduced from Ref. [9] with permission from American Chemical Society, Copyright (2011).) (c) Illustration of an idealized distance pattern of defects of 10 nm. (From Ref. [2] with permission from Wiley-VCH Verlag GmbH & Co, Copyright 2014.)

Since the quality of graphene is generally heterogeneous within a given sample, it is not sufficient to measure only one single spot spectrum of graphene. Therefore, scanning Raman microscopy was introduced, which allows the determination of the quality of graphene within a complete sample by applying statistical methods [7, 8]. Graphene, placed on a surface such as SiO_2, allows scanning Raman spectroscopy. Generally, silicon with 300 nm thick grown SiO_2 is used. These substrates make effective Raman measurements possible and allow visualizing graphene optically [11]. Using an optical microscope makes it also possible to distinguish between graphene, bi-layer graphene, and few-layer graphene by the blue contrast. Thus, films of SiO_2/G_1 can be used for statistical Raman microscopy (SRM). Generally, an SRM increment is used on the micrometer-scale. The useful increment depends on the size of the sample or the size of the flakes of graphene. Moreover, SRM can be used for Raman maps displaying properties of a given sample, reflecting a microscopic technique. Therefore, a small increment of scanning is used for high resolution to visualize inhomogeneity locally. After measurement, the I_D/I_G ratio of each spectrum and the Γ values of the D, G, and 2D peak are evaluated. With this information on hand it is possible to illustrate the quality of a given sample of graphene. There are several useful representations for illustrating the quality of graphene and one of them is the plot of the I_D/I_G ratio against Γ_{2D} as depicted in Figure 2.4 [12, 13].

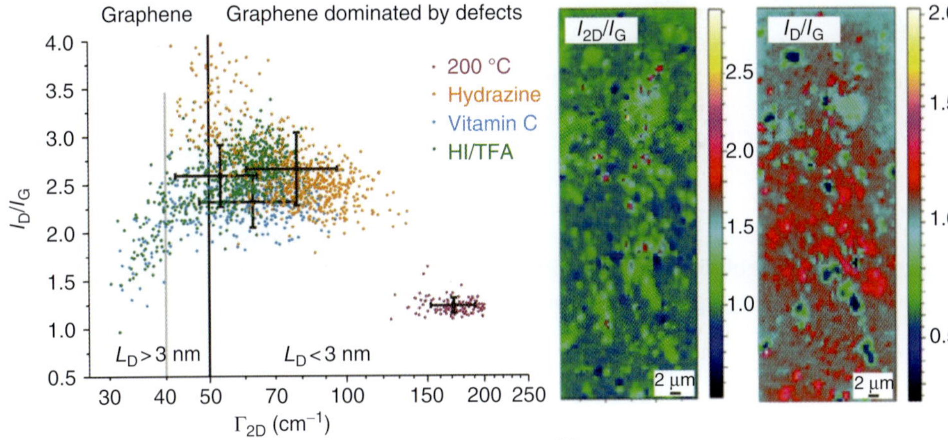

Figure 2.4 (a) Plotted data of the statistical Raman microscopic (SRM) analysis of films of graphene with different densities of defects; I_D/I_G versus Γ_{2D} (full-width at half-maximum (FWHM) of the 2D peak). Graphene derived from graphene oxide using different reducing agents. [12] – published by The Royal Society of Chemistry; (b) SRM maps of multilayered films of functionalized graphene derived from the reaction of C_8K (a donor graphite intercalation compound) and 4-*tert*-butylphenyldiazonium tetrafluoroborate. Local variations of the functionalization are visualized by plotting I_{2D}/I_G or I_D/I_G versus *x,y*-positions. (Adapted from Ref. [13] with permission from Macmillan Publishers Ltd: Nature Chemistry, Copyright 2011.)

2.4
Epitaxial Growth of Graphene from SiC

A common material used in high-power electronics is silicon carbide (SiC). The graphitization of hexagonal SiC was already reported in 1961 [14]. In the early stages of the growth of carbon layers from SiC, randomly stacked graphite was mainly obtained. Silicon sublimates from SiC at high temperature and layers of graphene or even graphite are left behind. In recent years, the method developed and the number of layers of graphene can be controlled by the process parameters. The quality of such graphene can be very high, with crystallites approaching hundreds of micrometers in size [15].

Drawbacks of this method are not only the high costs of SiC wafers but also the high temperatures of above 1000 °C, or even 1500 °C that are required. The high temperatures needed are related to the growth mechanism. Silicon from about three layers of SiC must sublimate from SiC to form one layer of graphene and, for the formation of few layers of graphene, even more layers are needed. It must be taken into account that the first carbon layer remains covalently bound to the underlying SiC and no sp^2-network of C-atoms is formed [16]. Therefore, a second carbon layer must be formed to generate a real layer of graphene. The first carbon layer can be seen as a buffer layer. Consequently, Si atoms must become mobile and because they cannot diffuse through a graphene layer they must diffuse to defect sites for sublimation. With respect to diffusion of Si atoms the rate of growth slows down, having control over the formation of layers [17]. High temperatures up to 1500 °C are necessary to make Si atoms sufficiently mobile and to minimize defect formation. Moreover, the pressure must be adjusted [18]. It was shown that treating the covalently bound carbon buffer layer on SiC by H_2 at 700 °C reduces the covalent bonds to form quasi free-standing graphene and hydrogen-terminated SiC [19].

The Raman spectra of graphene grown from SiC are less straightforward to measure because there is an overlay of the D and G peak with signals originating from the SiC substrate (Figure 2.5b). However, a subtraction of the SiC signals is possible and the real spectra of graphene can be visualized [21].

Moreover, graphene grown on SiC was transferred by separating the graphene from the substrate using a process based on a wet-chemical reaction producing gaseous species, such as oxygen from hydrogen peroxide. The oxygen intercalates between the substrate and the graphene, which finally causes the delamination of graphene from the substrate. High-resolution transmission electron microscopy was performed on such transferred graphene layers and the honeycomb lattice was well visible, which demonstrates that high quality graphene can be produced from SiC [22].

The synthesis of graphene from SiC is possible and may indeed be compatible with processes in the electronic industry. However, it must be taken into account that the process conditions must be very well controlled to generate pure graphene or bi-layer graphene. Moreover, the size of perfect graphene patches may be limited. In addition, the cost of energy is high and overcoming this problem remains

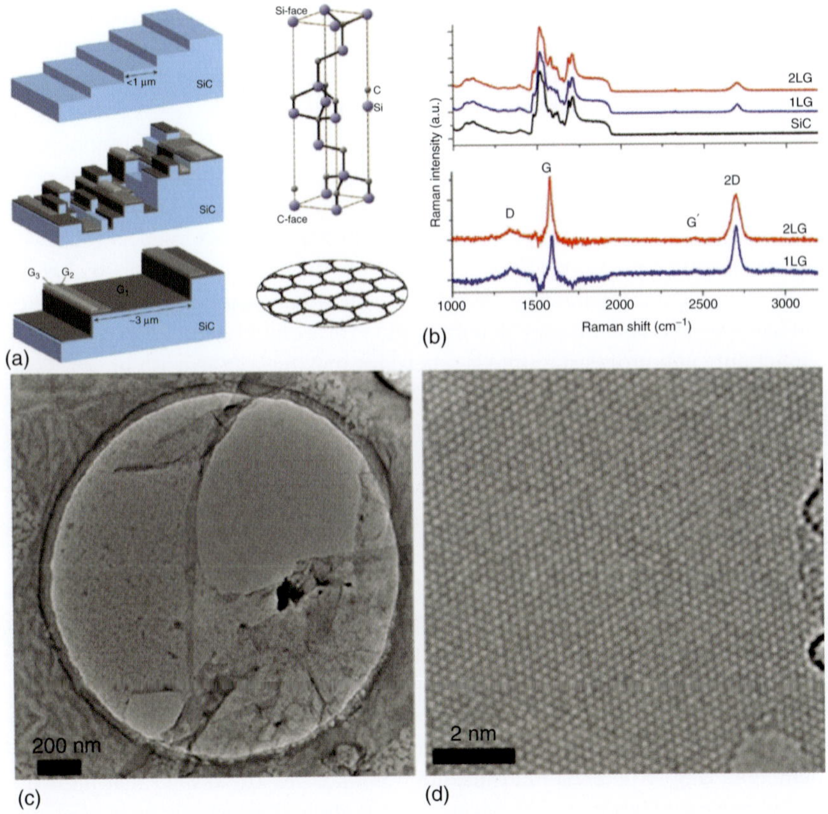

Figure 2.5 (a) Illustration of the growth of graphene on SiC. (Reproduced from Ref. [20] with permission from Macmillan Publishers Ltd: Nature Chemistry, Copyright 2009.) (b) Top: Raman spectra of graphene and bi-layer graphene on SiC, and bottom: spectra obtained after SiC-background spectra subtraction, shown for graphene and bi-layer graphene [21]. (c) High resolution transmission electron microscope image of graphene (overview) and (d) magnification that displays the intact honeycomb lattice. (Reproduced from Ref. [22] with permission from The Royal Society of Chemistry.)

challenging. Nevertheless, from the fundamental aspect of the study of graphene, bi-layer graphene, the formation of defects, and strain or dislocations, SiC grown graphene and bi-layer graphene remains a very likely preparation method [23, 24].

2.5
Graphene by Chemical-Vapor-Deposition

CVD is one of the most popular methods to provide graphene on surfaces. There are many reports on synthetic production conditions and it turns out that there are some general considerations that have to be taken into account to produce graphene on surfaces of high quality. Moreover, there are no generalizing reaction conditions available to predict the exact quality of graphene. Nevertheless, graphene growth catalyzed on metal surfaces from small molecules is a non-wet-chemical method to yield a high quality of graphene [25, 26].

Although there are some general obstacles in the preparation of graphene by the CVD approach, there are also some reports that provide methods to overcome the general problems of CVD graphene production. The number of layers of graphene can strongly vary depending on the metal surface used as well as on the preparation conditions [26, 27]. In addition, the growth mechanism changes upon changing the metal. A typical substrate for graphene growth is nickel, which can form nickel carbide at growth temperatures of around 1000 °C. The processing temperature and processing time is crucial for making graphene on nickel. Typical conditions involve a pretreatment of the nickel surface at 1000 °C in hydrogen for a period of about 1 h. Different types of carbon sources can be used, such as methane, acetylene, or other gases. For methane, a 20 min growth is facilitated with a mixture of argon, hydrogen, and methane. Using this method yields not only graphene, but also few-layers. It is interesting to note that typical cooling rates range between 0.1 and 20 $°C\,s^{-1}$. Upon cooling, nickel carbide decomposes and carbon segregates on the surface and forms graphene. However, the growth conditions must be well controlled to avoid producing excessive amounts of graphite.

Beneath the preparation of few-layer graphene there are also grains in polycrystalline nickel and they determine the size of the single crystalline graphene. When crystals of graphene grow, they will interact with each other and start to form continuous films. A general problem in this regard is the formation of non-six-membered ring structures and thus, grain boundaries are formed that limit the quality of films of graphene. The problem of grain boundaries not only exists for the growth of graphene on nickel, but also on other metals. Even single crystals of nickel have been used for the growth of graphene and it was found that much smoother layers of graphene can be obtained [28] (similar process, as depicted in Figure 2.6).

It remains challenging to use nickel as the substrate for graphene production due to the preferred formation of few-layer graphene. Thus, other metals are currently

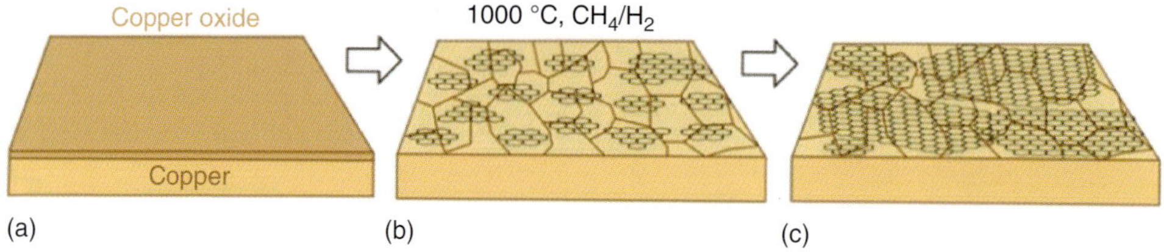

Figure 2.6 Illustration of the formation process of graphene grown on a catalytically active metal surface, such as copper; (a) copper substrate with oxides on top; (b) growth of islands of graphene and grains illustrated; (c) patches of graphene merging forming grain boundaries. (Reproduced from Ref. [25] with permission from The Royal Society of Chemistry.)

more favored for the production of graphene. Many different metal surfaces, such as copper, platinum, ruthenium, iridium, cobalt, rhenium, or germanium have been used. However, the most popular substrate is still copper, since the growth parameters can be adjusted to avoid the growth of few-layer graphene. Typical copper foils used are polycrystalline with a thickness of around 25 μm. The growth temperature is about 1000 °C, and a carbon source, such as methane, is used in a mixture with hydrogen. While the carbon source dissolves in nickel, the formation mechanism of graphene on copper differs [25, 26]. Copper does not form carbides and carbon can only barely, if at all, dissolve in copper. The carbon source, such as methane, reacts on the surface and the C–H bonds are cleaved. However, a direct interaction of methane with copper does not seem to be plausible and the adsorption of CH_x species on copper is not feasible as well. In contrast to the direct interaction of methane with copper, the attention was turned to CuO_x species present on the surface of copper [29]. Copper forms species such as copper-I-oxide and under growth conditions the CuO_x species are mobile on the surface. It was suggested that these oxo-species have a strong influence on the nucleation and growth of graphene on copper. In addition, it was demonstrated that the oxygen content in copper can indeed influence the kinetics of the growth of graphene and that its content in copper is a critical parameter for that control. Polycrystalline copper exhibits more active grains that can have a stronger interaction with oxygen compared to other surface atoms. Thus, adsorption or chemisorption of oxygen atoms can passivate the grains of copper. Indeed, the amount of nucleation centers could be reduced by pretreating copper with oxygen. The nucleation centers were dramatically reduced by oxygen treatment of the copper foils and 0.01 nucleation sites per square millimeter was found [30]. This finding resulted in a synthetic strategy allowing the growth of single crystals of graphene with diameters of about 1 cm. Moreover, it was reported that the growth of graphene changed from hexagonal structures to dendritic structures and the growth was reported to accelerate due to oxygen species. It was found that the activation energy for the growth was reduced from 1.76 to 0.92 eV. It can be expected that C–H dissociation is accelerated and CH_x species can be generated at lower activation barriers. Thus, generated CH_x species can easily be formed and diffuse rapidly to form C–C bonds. Especially, the finding that very large crystals of graphene can be formed strengthens the idea that the grain of the copper foil plays only a minor role. However, the crucial drawback of CVD growth of graphene on copper is the different thermal expansion coefficient of graphene and copper that causes wrinkling of graphene after cooling. These wrinkles also reduce the quality of the produced graphene (Figure 2.7).

Another process and synthetic strategy for the growth of graphene was reported, which involves germanium as substrate [31]. Drawbacks such as the wrinkling of graphene upon cooling and grain boundaries that evolve when crystals of graphene merge have been reported to be solved. Overcoming these crucial problems of graphene growth are breakthroughs in the technical production of graphene on surfaces. However, it must be kept in mind that the exact processing parameters are very critical in CVD growth and although a process

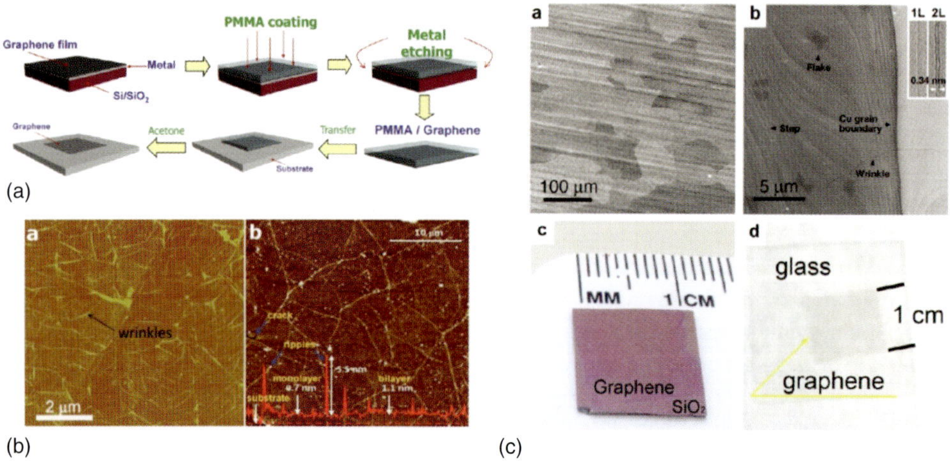

Figure 2.7 (a) Illustration of the transfer of graphene from the metal catalyst by PMMA coating, metal etching, transfer on the desired substrate, followed by dissolution of the polymer, leaving graphene back on the substrate. (Reprinted from Ref. [26] with permission from American Chemical Society, Copyright 2013.) (b) AFM images of graphene films transferred onto SiO_2, displaying cracks and wrinkles. (Reproduced from Ref. [25] with permission from The Royal Society of Chemistry.) (c) (A): image of graphene grown on a foil of copper and visualized by scanning electron microscopy (SEM); (B): magnification SEM image of graphene on copper with wrinkles, grains, and steps; (C): transferred graphene on isolating SiO_2/Si wafer; (D): transferred graphene on a glass slide. (Reprinted from Ref. [26] with permission from the American Chemical Society, Copyright 2013.)

is described in detail, its reproducibility on other growth systems may need long optimization. Consequently, reproducibility and full control of the process is a technological question.

The common strategy to grow graphene is based on the formation of isotropic nucleation sites that results in defective grain boundaries. The other one is based on minimizing nucleation sites to grow single crystals of graphene. However, this process is rather slow and maintaining a low density of nucleation sites is challenging. The novel strategy followed on germanium (110) is based on anisotropic nucleation sites. This implies that graphene on germanium (110) starts to grow in a favored direction. When all initial graphene crystals start to grow in the same direction, the edges of graphene will fit perfectly when crystals start to merge and therefore, grain boundaries are avoided. While graphene growth on germanium (111) results in isotropic growth, the anisotropic growth was realized on germanium (110). In addition, the germanium used was terminated by hydrogen, which minimizes the interaction of germanium and graphene. Lowering the interaction forces make a transfer easier. Another benefit is due to the ability to grow germanium (110) on silicon wafers, which minimizes costs. This approach also facilitates transfer of grown graphene to other substrates. Even the used germanium wafer was recycled for another growing process. Since the thermal expansion coefficients of germanium and graphene are very similar, a rather smooth surface of graphene was yielded. The growth was accomplished between 900 and 930 °C with methane as carbon source, diluted with 1–2 % of hydrogen.

Although growth methods for the synthesis of graphene have been developed, catalytically active metal surfaces are necessary. However, the desired substrate is mostly not a metal surface, but an insulating surface, such as SiO_2 or a polymer/transparent polymer, targeting device fabrication. Therefore, it is necessary to transfer the grown graphene from the metal surface to another one. The strategies most often used are either based on etching or based on using a polymer for transfer. It is well known that graphene on, for example, copper can be suspended on water using, for example, iron-III-chloride to solute the metal. After the copper dissolves, graphene floats on the surface of water and can be transferred onto another substrate, such as a silicon wafer with a thermally grown silicon dioxide surface. Another method uses a polymer film, such as poly-methylmethacrylate (PMMA, Figure 2.7). The graphene binds to the PMMA and can be transferred onto the desired substrate. Subsequently, the PMMA is dissolved with acetone to leave the graphene on the surface. However, there are some drawbacks in both methods. The etching process is problematic, because heavy metals must be dissolved, which is not environmentally friendly. Moreover, the process is not atomically economic and metal impurities remain on the graphene as contamination. The PMMA approach leaves polymeric residues on the graphene and removing them is almost impossible. Another transfer method was reported by using a thermal release tape that is able to adsorb graphene and release it by heating to 100 °C [32]. This approach was also used for roll-to-roll production of graphene.

The quality of CVD graphene is high enough for technical applications, even with grain boundaries and therefore, it is a technical question if costs, quality, and scaling of the process can meet the demands necessary for applications, such as transparent and conductive coatings or others. However, for future applications, such as spin electronics or ultra-high frequency applications, the requirements of the quality of graphene are absolute, with no defects at all [33].

2.6 Delamination of Graphene from Graphite

Natural graphite is available in large and almost infinite quantities. The fact that a single layer of graphite is graphene encourages scientists to use graphite as the initial material to make graphene in large quantities. Surely, it is illusive to make continuous films of graphene from graphite because of the flake limit. At best, films of flakes of graphene can be expected. Nevertheless, for basic research and applications where diameters on the nanometer- or micrometer-scale are needed, graphene from graphite is the method of choice. In the following sections, mechanical methods for the production will be highlighted. The advantages and disadvantages of the approaches will also be discussed.

2.6.1
Mechanical Cleavage of Graphite

The most prominent method to make graphene was introduced by Novoselov *et al.* [34]. They not only surprisingly succeeded in the isolation of graphene, but also placed it on a SiO_2/Si substrate that allowed the evaluation of the outstanding physical properties of graphene. The method introduced is called the "scotch tape" method. An adhesive tape is used for the exfoliation of graphite down to the single layer of graphene. The procedure involves repeated pealing of few-layers of graphene from flakes of graphite (Figure 2.8). Then, the adhesive tape is placed on the SiO_2/Si wafer and after gently pressing and slowly peeling off the adhesive tape from the wafer it is possible to deposit some flakes of graphene on the surface of the wafer. Flakes of graphene with lateral dimensions of several micrometers can be obtained and identified as graphene by Raman spectroscopy. It should be noted that the source of graphite is important and the right one must be identified with experience. Among graphene, bi-layer graphene and few-layer graphene are also deposited. However, bi-layer graphene and tri-layer graphene are also of interest in research [36, 37]. Graphene can be identified by a light blue contrast visible in an optical microscope if placed on 300 nm SiO_2/Si wafer [11]. However, it is mandatory to identify graphene as a single layer and the quality of graphene must be evaluated also by Raman spectroscopy [6]. The advantage of peeling graphene off from graphite is due to its high quality. No grain boundaries are present and the lattice is free from defects within a very large area up to the micrometer-scale. Thus, this

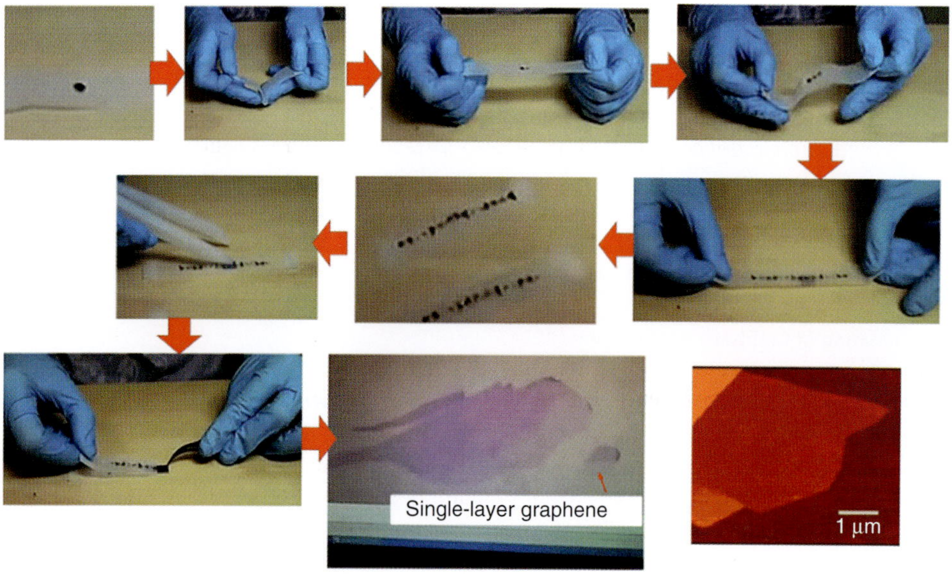

Figure 2.8 Mechanical cleavage of graphene from graphite using adhesive tape. Graphene is deposited on to SiO_2/Si wafers and identified by optical microscopy and AFM. (Reproduced from Ref. [35] with permission from The Royal Society of Chemistry.)

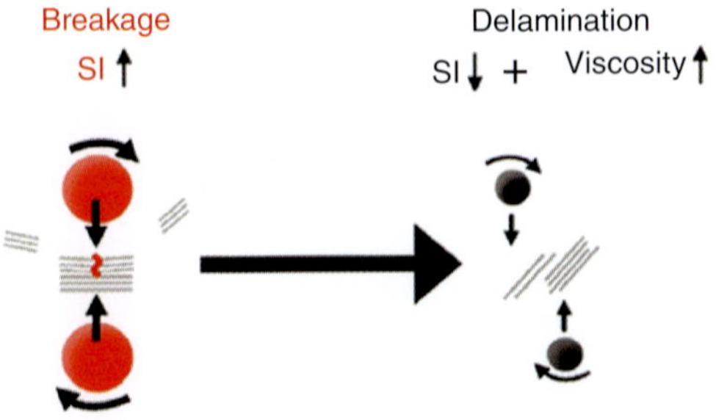

Figure 2.9 Illustration of the adaption of processing parameters, such as stress intensity or viscosity of the solvent, using also beads of different size. Optimization of the yield of flakes with a reasonable lateral dimension of hundreds of nanometer of graphene and few-layer graphene is possible. Adaption of processing parameters improve the delamination efficiency and reduce the formation of defects within the carbon framework. (Reprinted from Ref. [38] with permission from Springer Science+Business Media, Copyright 2015.)

method of preparing flakes of graphene is favored for basic research to evaluate the fundamental properties of graphene. However, for technical applications, this method is not efficient and therefore, other methods have been developed that are scalable (Figure 2.9).

2.6.2
Liquid Phase Exfoliation of Graphite – Stirred Media Mills

Scalable methods to prepare graphene are rare [39]. However, a facilitated approach is performed using ball milling, a grinding method with graphite as carbon source. Beads that might consist of either polystyrene or denser oxides such as zirconoxide are typical grinding balls [40]. Hurdles in this approach are, on the one hand, avoiding breaking of flakes and, on the other hand, avoiding aggregation. Thus, processing parameters must be well controlled and a stabilizer must be used.

Nonionic surfactants, such as TWEEN® 80 that bear polyether groups are used in an advanced approach to prepare graphene and few-layer graphene [38]. The stress intensity and specific energy input introduced by the balls must be controlled to minimize defect formation and to increase the yield of few-layer graphene and graphene. Recently, it was demonstrated that a low-specific energy input preferentially yields few-layer graphene, while high-energy input facilitates breaking of flakes to produce graphite nanoparticels accompanied with defect formation. Moreover, the solvent viscosity plays a crucial role in the delamination of graphene flakes from graphite. It was shown that increasing the viscosity from 1 to 6 mPa · s enhances the delamination by shear and friction and reduces in-plane fracture. The dampening of beads in a more viscous medium was suggested as reason for the more efficient delamination of graphite to graphene and few-layer graphene. However, even if graphene was not exclusively produced,

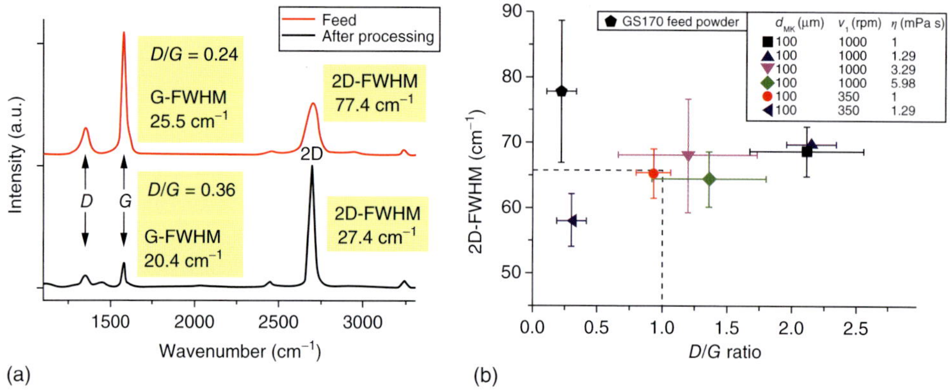

Figure 2.10 (a) Raman spectra of graphite (upper spectrum) and graphene (lower spectrum). The FWHM ($27.4\,cm^{-1}$) of the 2D peak is characteristic for graphene. (b) Analysis of SRM data illustrating the quality of graphene and few-layer graphene. The quality of the product increased with improving the processing parameter for the stirred media milling process. Parameters such as, size of beads, viscosity or energy input were optimized. (Reprinted from Ref. [38] with permission from Springer Science+Business Media, Copyright 2015.)

it must be kept in mind that few-layer graphene and graphene can be produced in large quantities. In addition, scale up to the liter-scale was demonstrated without influencing the quality of graphene and few-layer graphene. The product obtained by such methods is heterogeneous and thus, it is important to analyze the number of layers of graphene as well as the quality of graphene and few-layer graphene. As outlined earlier, statistical Raman microscopy (SRM) is a suitable method to gain information about the density of defects (quality) by the interpretation of the I_D/I_G ratio, as well as by analyzing the FWHM of peaks, such as the 2D peak at about $2700\,cm^{-1}$. In Figure 2.10 typical Raman spectra of graphite and delaminated graphene are shown. The overall quality of graphene, which is dependent on processing parameters, is screened. A reasonable quality of few-layer graphene was achieved, characterized by the FWHM of the 2D peak between 55 and $60\,cm^{-1}$ with an I_D/I_G ratio of about 0.3 only.

A survey on ball-milling of graphite in solvents, as a method to prepare graphene and few-layer graphene (FLG), was presented in great detail [40]. The few-layer graphene yield was thoroughly analyzed by SRM. Surfactant-free FLG suspensions were prepared by wet-media delamination of graphite in organic solvents. Among the investigated solvents, the highest FLG concentration was found for NMP in agreement with the lowest van der Waals interaction energy. The graphite delamination process in NMP was optimized by varying the delamination tool, the delamination media size, and the stirrer rotation speed. Processing of graphite in NMP using a stirred media mill under optimized process conditions ($100\,\mu m$ ZrO_2 beads, stirrer tip speed: $3.4\,m \cdot s^{-1}$) yields a suspension of few-layer graphene that is nearly free from graphitic impurities. The few-layer graphene production rate and specific production rate is $73\,mg \cdot h^{-1}$ and $0.43\,g\,l^{-1} \cdot h^{-1}$, respectively, which is higher when compared to ultrasound-assisted graphite delamination (about $1-33\,mg \cdot h^{-1}$ and $2.6 \times 10^{-3}-0.33\,g\,l^{-1} \cdot h^{-1}$, respectively) at comparable energy costs per gram of produced few-layer graphene. Thus, also from the economical point of, presented method is interesting for developing an industrial few-layer graphene production process. The presented parameter studies can be also regarded as a guideline for process design. As expected, the percentage of few-layer graphene in the products derived from SRM agrees well with the percentage of few-layer graphene determined from particle thickness measured by AFM. Therefore, the distribution of Γ_{2D}, which is much easier and faster to measure than an AFM flake height statistics, also gives information about the number of layers per particle. Evaluation of the Γ_{2D} distributions reveals that produced few-layer graphene consists of mainly tri-layers.

2.6.3
Liquid Phase Exfoliation of Graphite – Sonication

There are manifold examples of sonicating graphite in either organic solvents or mixtures of water with surfactants [39]. However, in all examples, few-layer graphenes are yielded and graphene is only produced in small amounts. Moreover, the sonication of graphite in solvents is hard to scale up and long periods of time amounting to weeks are necessary for reasonable results [41]. Thus, a growing content of graphite nanoparticles among nanometer-sized few-layer graphene is difficult to avoid.

It was demonstrated that few-layer graphenes with flake diameters in the range of 100–500 nm are formed in quite large portions with a minor fraction of graphene. That is due to breaking of graphite flakes as a concurrent process to the delamination process. Density gradient ultracentrifugation is a method suitable for the analysis of the number of graphene layers of sonicated samples (Figure 2.11). A major part of such samples consists of few-layer graphene and a certain fraction of real graphene in minor quantities can be identified [43]. Raman spectra of isolated fractions of graphene prove the presence of the number of layers [42].

Typical surfactants used for exfoliation are depicted in Figure 2.12. They are generally based on the combination of a π-system and other functional groups that provide water solubility. In addition, polymers were used yielding water dispersed few-layer graphene after a first exfoliation step in NMP or *o*-dichloro benzene [39]. The disadvantage of surfactants is that they are persistent on the surface of graphene and few-layer graphene and therefore, attention must be paid to the type of surfactant used, which may be either a disadvantage or a benefit for the desired application.

Solvents of high boiling points were found to be suitable for the exfoliation of graphite with NMP, *ortho*-dichloro benzene (*o*-DCB) or dimethylformamide (DMF) as the most prominent ones. In addition, fluorinated aromatic molecules, such as pentafluorobenzonitrile, were used. A feature these solvents have in

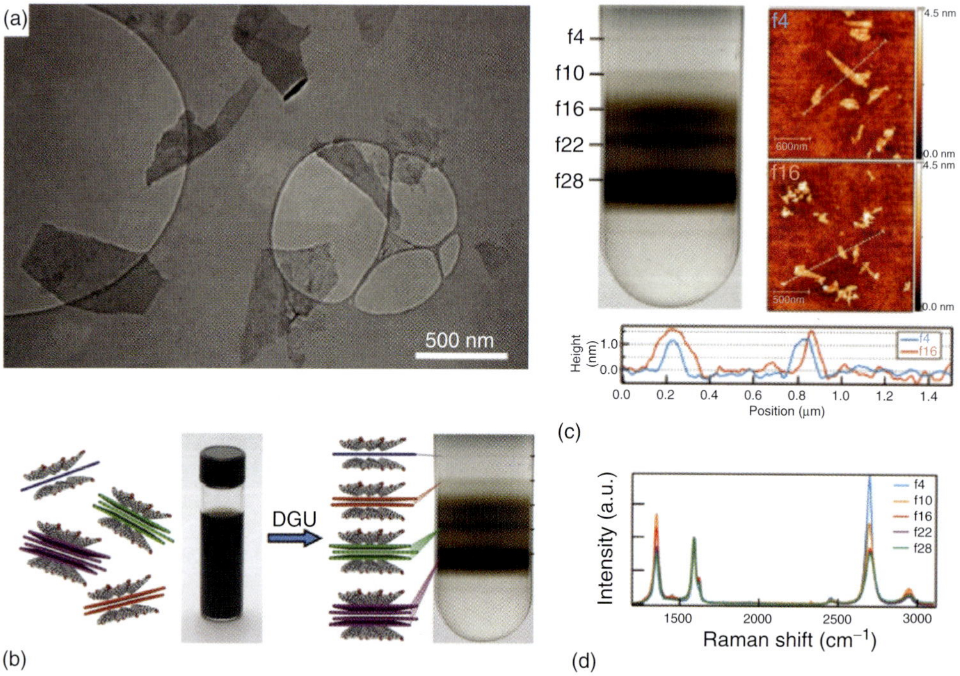

Figure 2.11 (a) TEM image of graphene and few-layer graphene after sonication of graphite in *N*-methylpyrrolidone for 312 h. (Reprinted from Ref. [41] with permission from the American Chemical Society. Copyright 2011.) (b) Polydisperse surfactant-stabilized dispersion of graphene and few-layer graphene. Density gradient ultracentrifugation (DGU) yielded fractions of graphene, bilayer, and few-layer graphene. (c) Fractions of stabilized graphene, separated by DGU, with related AFM images and height profiles. (d) Raman spectra of graphene fractions prepared by DGU, F4 fraction relates to graphene. (Reprinted from Ref. [42] with permission from the American Chemical Society, Copyright 2009.)

common is that their surface tensions are close to $40\,\mathrm{mJ \cdot m^{-2}}$ meeting that of graphite and graphene . Therefore, they should be the best candidates for exfoliating graphite. A hurdle in using NMP or DMF is their toxicity toward organs. Numerous attempts have been made to produce graphene of high concentration. However, up to now, the commonality among these methods is that few-layer graphene with flakes sizes <1 μm are yielded among nanometer-sized graphite.

2.6.4 Liquid Phase Exfoliation of Graphite – Shear Mixing

Shear-forces, using a shear mixer are an alternative and scalable method to prepare few-layer graphene and graphene [44]. As shown in this example, shear mixing was used also in combination with surfactant stabilization. In addition,

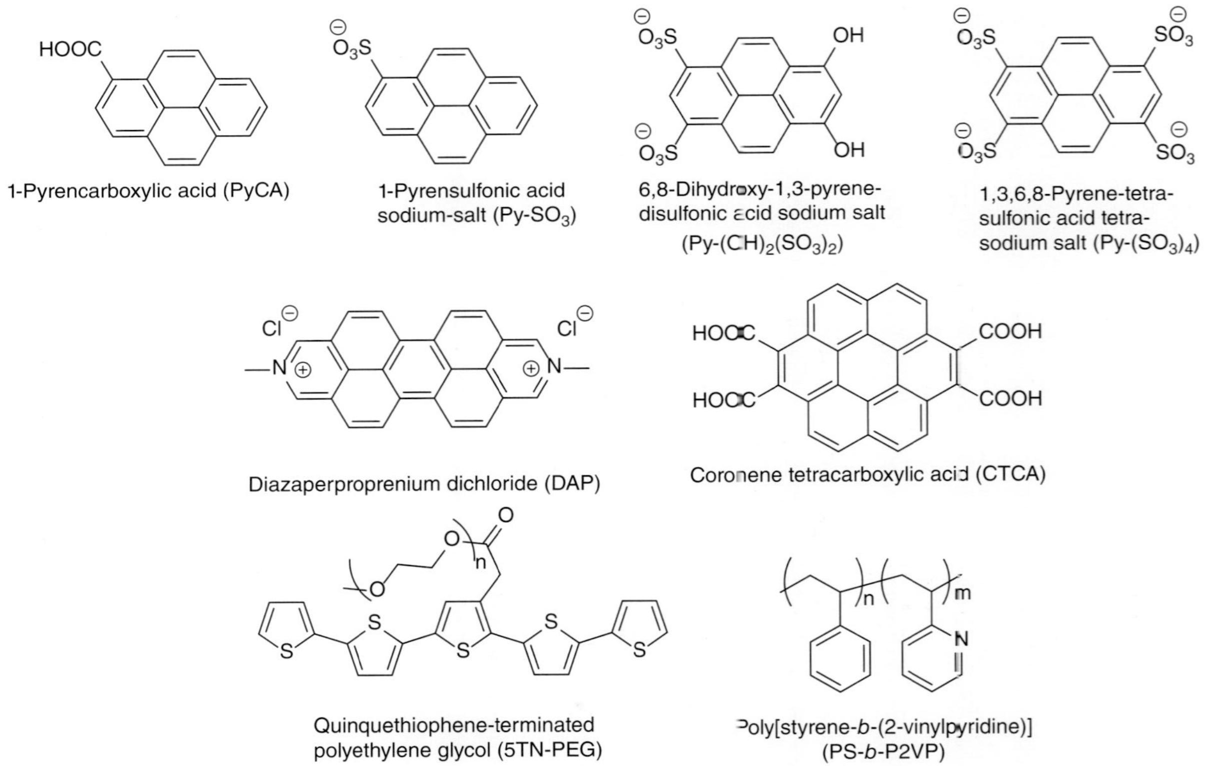

Figure 2.12 Selection of surfactants suitable to stabilize graphene and few-layer graphene in aqueous solution [39].

exfoliated graphene and few-layer graphene dispersions were produced in NMP. Although most flakes are few-layer graphene and the lateral dimension of flakes is smaller than 1 μm it is one of the rare examples by which Raman spectroscopy results prove that graphene is yielded by a sharp 2D peak as depicted in Figure 2.13. Nevertheless, other Raman spectra clearly indicate few-layer graphene nature by the broad 2D peak.

2.6.5 Liquid Phase Exfoliation of Graphite Using Smart Surfactants

Another non-covalent approach to graphene is facilitated using a smart surfactant. In this selected example, a bolaamphiphile is used to delaminate the layers of graphene from graphite [45, 46]. Flakes of natural graphite are used as the carbon source, and graphene together with few-layer graphene were obtained using the water-soluble perylene as depicted in Figure 2.14. The π-system of the perylene can interact with the surface of graphene and the carboxylic acids attached to the perylene provide water-solubility. The flakes of graphene can be identified by Raman spectroscopy. Graphene with a flake size of about 1 μm and a moderate defect density of approximately 0.01% was found, as indicated by the D peak. The presence of a moderate amount of defects may be a prerequisite for the successful delamination. An SRM analysis was also performed, which revealed the polydisperse nature of the sample.

2.6.6 Electrochemical Exfoliation of Graphite

The electrochemical exfoliation of graphite electrodes is a scalable method to produce few-layer graphene and it was demonstrated that the density of defects in the generated material is rather low, by an I_D/I_G ratio of 0.4 [47]. The aim of the work is to produce solution-processable graphene to enable electronic applications. Mainly, about 80% of flakes of G_{1-3} with a C/O ratio of 12.3 were produced by this electrochemical approach. The proposed mechanism of exfoliation that involves the electrochemical generation of reactive oxygen species that can oxidize graphite is shown in Figure 2.15. This oxidation leads to an expansion of layers and sulfate ions can penetrate the interlayer space. Finally, graphene and few-layer graphene that is suitable for further processing is obtained. AFM images reveal graphene with lateral dimensions of about 5 μm. It was demonstrated that large area and highly conductive films of overlapping flakes of graphene can be produced on various substrates and field-effect transistors were produced with a performance of several tens of centimeters squared per volt-second. However, these values are still at least two orders of magnitude lower compared to those of graphene. Further development of a controlled electrochemical approach may not only be scalable, but can also produce graphene as a major species.

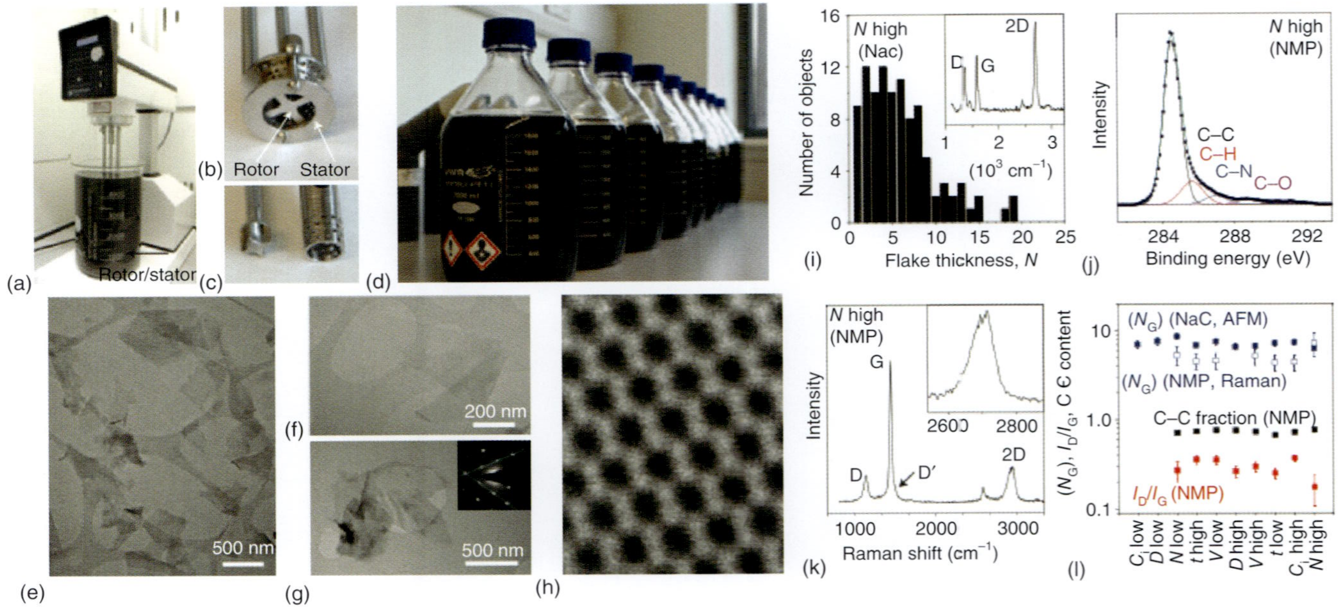

Figure 2.13 Preparation of graphene and few-layer graphene by shear mixing. Graphene can be identified by Raman spectra even if graphene is a minor fraction among few-layer graphene. X-ray photoelectron spectroscopy can identify C–N bonds originating from NMP solvent. HR-TEM images reveal areas of an intact honeycomb lattice. (Reprinted from Ref. [44] with permission from Macmillan Publishers Ltd: Nature Materials, Copyright 2014.)

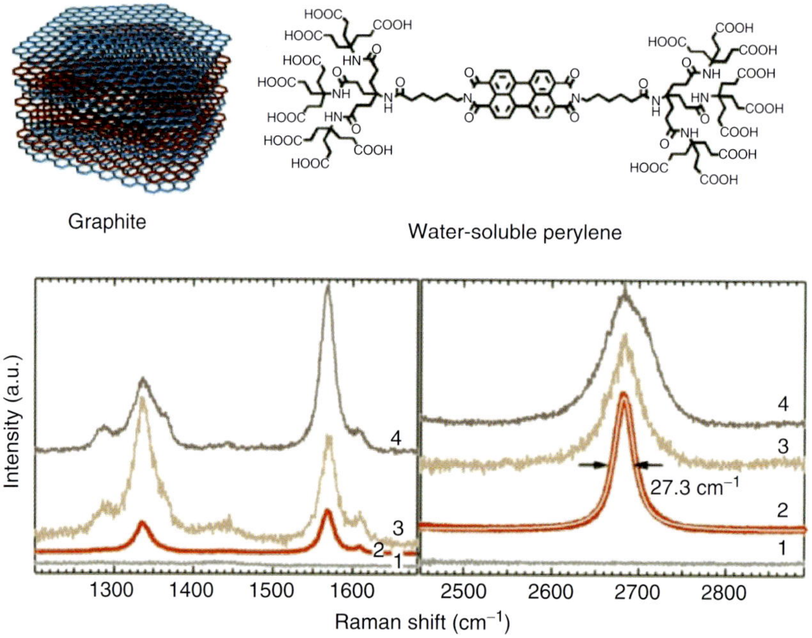

Figure 2.14 The water soluble bolaamphiphile, comprising a perylene moiety as well as carboxylic acids, was used to delaminate graphene from graphite in water without the need of sonication or shear forces. (From Ref. [2] with permission from Wiley-VCH Verlag GmbH & Co, Copyright 2014, and Ref. [45], Copyright 2009.)

2.7 Wet-Chemical Functionalization and Defunctionalization

2.7.1 Reductive Functionalization of Graphene

The chemical reduction of graphene, by means of transferring electrons into the conduction band of graphene, is an activation step for the functionalization with electrophiles [13]. Negatively charged graphene, called graphenide, is present in donor graphite intercalation compounds (GICs). The alkali metal in intercalated graphite acts as the electron donor [48, 49]. Stable GICs consist of subunits of C_6Li or C_8K. In addition, C_2Li is known but can only be formed under high pressure [50]. Donor-GICs can be dispersed in solvents such as dimethoxyethane to form few-layers and single layers of graphenide [13, 51]. These graphenides can subsequently undergo addition reactions with electrophiles like alkylhalides or arylhalides, as illustrated in Figure 2.16. Following this functionalization route, it is possible to functionalize even individual layers of graphene by a C–C coupling reaction. It was demonstrated that graphene derivatives, functionalized by C–C coupling on the surface, are thermally stable in a wide range, as outlined below.

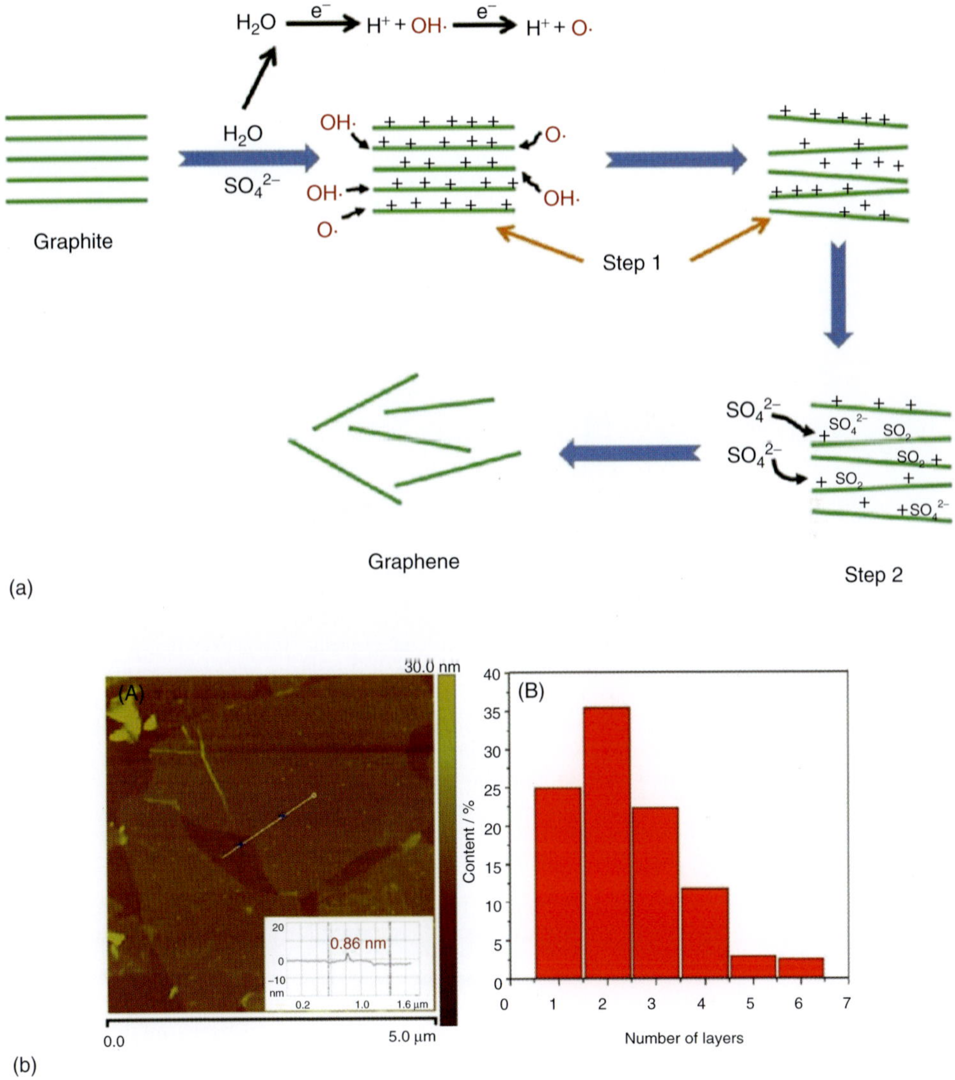

Figure 2.15 (a) Illustration of the proposed mechanism of electrochemical exfoliation, involving the generation of reactive oxygen species that lead to partial exfoliation of the carbon layers, followed by sulfate penetration of layers and delamination to graphene and few-layer graphene. (b) AFM image of graphene and statistical analysis of the number of layers. (Reprinted from Ref. [47] with permission from the American Chemical Society, Copyright 2013.)

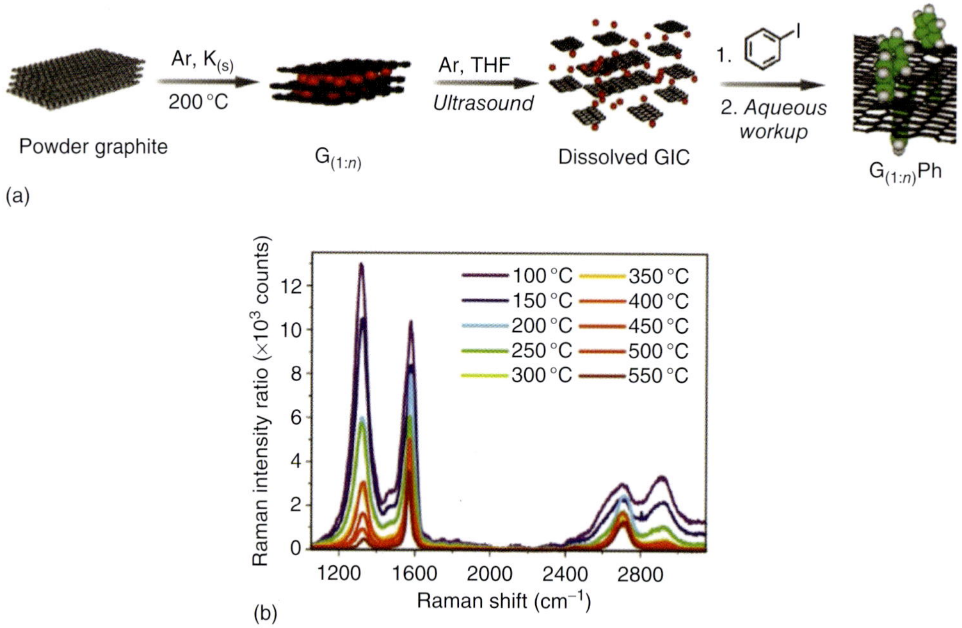

Figure 2.16 (a) Illustration of the functionalization route producing a graphite intercalation compound, dispersing it in a solvent to generate graphenide, followed by addition of an electrophile, here aryliodide. (b) Thermal defunctionalization of addends forms stacks of graphene. The D peak that is induced by defects (here due to phenyl-addends) is thermally removed. (From Ref. [52] with permission from Wiley-VCH Verlag GmbH & Co, Copyright 2014.)

An example of this functionalization route is the reaction of graphenide with 4-*tert*-butylphenyldiazonium tetrafluoroborate in dimethoxyethane, a reaction that was demonstrated to form arylated graphene [13]. A similar functionalization of a graphenide was achieved with *n*-hexyl iodide as depicted in Figure 2.17a [53]. Raman spectroscopic analysis of hexylated graphene proves the single layer nature. As outlined earlier, Γ_{2D} values <40 cm^{-1} is an indication for the single layer nature. In this example, the 2D peak remains sharp after functionalization with an I_D/I_G ratio of 2 that can be related to a degree of functionalization of about 0.2% [7].

Another example is the reversible hydrogenation of graphenide, as shown in Figure 2.17b [54]. Protons are the smallest electrophiles and the complete functionalization of graphene leads to graphane. However, it remains extremely challenging to fully hydrogenate graphene. Approaches toward graphane are summarized in the literature [55]. It was demonstrated on individual layers of graphene that hydrogenation is indeed possible and thermally reversible [56].

The reductive functionalization route of graphene with graphenide as a reactive intermediate reflects a growing field toward functionalized graphene. Especially, C–H and C–C bond formation are possible and the latter reflects a precursor to

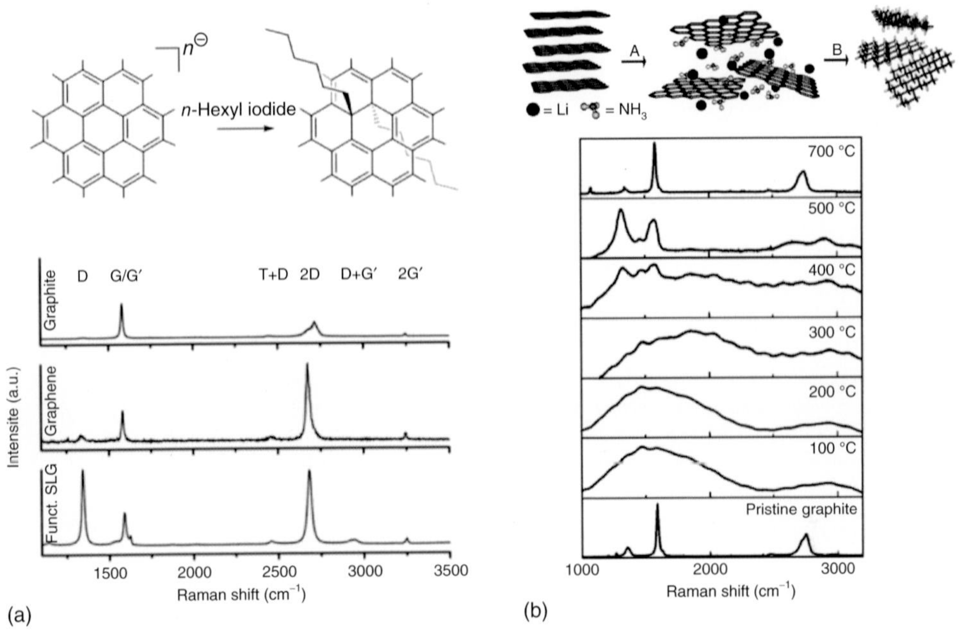

Figure 2.17 (a) Top: Reaction of graphenide with the electrophile hexyl iodide to hexyl-functionalized graphene. (From Ref. [2] with permission from Wiley-VCH Verlag GmbH & Co, Copyright 2014.) Bottom: Raman spectra of graphite, graphene, and functionalized graphene. (Reproduced from Ref. [53] with permission from The Royal Society of Chemistry.) (b) Top: Wet-chemical hydrogenation of graphenide by protons to form fluorescent partially hydrogenated graphene. Bottom: Temperature dependent Raman spectroscopy that depicts the reversibility of the hydrogenation by heating the sample to 700 °C. (From Ref. [54] with permission from Wiley-VCH Verlag GmbH & Co, Copyright 2013.)

even more complex graphene derivatives. However, reaction conditions must be optimized and further developed to generate dispersions of functionalized single layers. At the current state of the art it cannot be excluded that restacked functionalized layers of graphene or functionalized few-layer graphenes and graphite are obtained in mixtures.

2.7.2
Oxidative Functionalization of Graphene

The concept of reductive functionalization facilitates reduction of layers of graphene, starting from graphite. In contrast to that approach, oxidative functionalization of graphene involves oxidation of graphite, either by electron removal or addition of oxygen atoms. Electrical oxidation facilitates the removal of electrons from the valence band of graphite and graphene, respectively. However, addition of oxygen to the carbon framework of graphene leads oxo-functionalization, a method that is most often used to individualize graphene

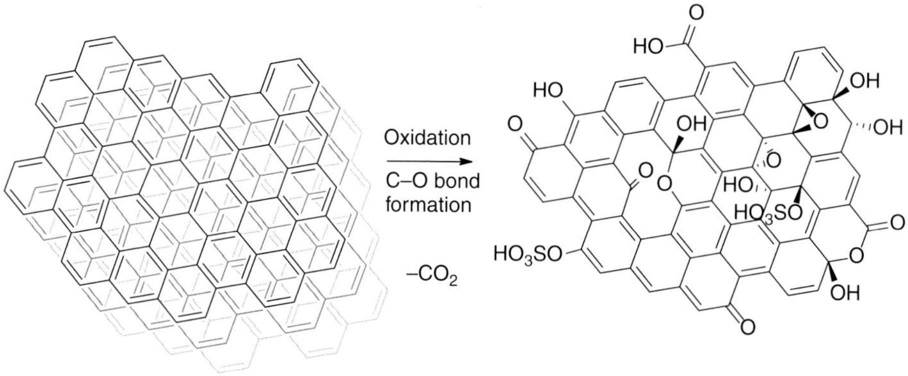

Figure 2.18 Chemical sketch for the oxidation of graphite to graphene oxide. (Reprinted from Ref. [58] with permission from Springer Science+Business Media, Copyright 2014.)

from graphite in a soluble form. These individual layers are called graphene oxide (GO) [57]. Stacks of GO, which are also the first oxidation product of graphite with the layers of functionalized graphene still connected, are called graphite oxide (GrO).

GO is a single layer of oxo-functionalized carbon. GO has a long history as GrO. GrO was also called graphitic acid [57]. The chemical functionalization of layers of graphene by oxygen functional groups generates sp^3-hybridized carbon. The hydrophilicity of oxo-functionalized graphene makes dispersions possible in polar solvents. In Figure 2.18, a chemical structure model of GO is shown. It must be noted that over-oxidation by means of CO_2 formation is dominant following common synthetic procedures [59]. In recent years, the loss of carbon was quantified to about 1 CO_2 on 35 carbon atoms, reflecting a density of defects of about 3%, following common oxidation procedures. Moreover, the loss of one carbon atoms results in the generation of three edge carbon atoms. It can be suggested that a C–O bond is formed at each edge and therefore, roughly 10% of oxygen can be added just on edges of defects within the basal plane of GO, assuming a density of defect of about 3%. A plausible chemical structure of edges of a defect is shown in Figure 2.18 with a ketone and a semi-acetal function. As outlined in further text, on average it is possible to avoid over-oxidation and it is possible to minimize the formation of defects to about 0.3–0.04%. The defect density can be determined by Raman spectroscopy of the graphene obtained from GO after chemical reduction.

GO is a versatile and processable derivative of graphene, with a variable density of defects. It can act as a precursor for graphene. However, with increasing the density of defects, the electronic properties become limited and do not reach those of perfect graphene, which requires a defect density as low as approximately 0.001% and below. At such a low density of defects Raman spectroscopy does not identify a significant intensity of the D peak and thus, the unusual physical properties of graphene could be revealed with precision [60]. Defects limit the electronic properties of graphene derived from GO, such as the charge carrier mobility values. Thus, defects must be considered as an inherent property of GO and the derived graphene. Up to now, the average density of defects of usually prepared GO is about 3%, see above, but the lowest defect density of graphene derived from flakes of GO exhibits a defect density of about 0.01% and therefore, roughly only one order of magnitude more defects than pristine graphene.

Furthermore, GO is a collective term for functionalized graphene with various oxygen functional groups. The contribution of oxo-functional groups is in arbitrary amounts and depends on the preparation conditions. Moreover, the defect density within the carbon framework is undefined. Within roughly the last 150 years, the structure of GO was analyzed thoroughly and different structure models have been developed [61, 62]. However, as synthetic chemistry on GO was developing, novel structure motives were found.

Although GO is a polydisperse material, applications such as transparent electronics [63, 64], charge storage [65], sensors or biosensors [66], composite materials [67], pollution management [68], and biomedical applications were demonstrated [69, 70].

2.7.2.1 Generalized Synthesis of GO

Graphite is generally used as carbon source for the preparation of GO. The synthetic protocols can be divided in four steps. However, these steps proceed in part simultaneously (Figure 2.19).

First, graphite is mixed with an acid, either nitric acid or sulfuric acid. Second, the oxidant is added in portions for oxidation. Under these conditions an acceptor type GIC is formed and with respect to the used acid graphite nitrate and graphite sulfate are formed as intermediates, respectively [49]. Oxidation and functionalization of graphite is accompanied by the intercalation process that proceeds with the addition of oxidant. Potassium chlorate or potassium permanganate is generally used. Chlorine dioxide from chlorate and manganese heptoxide from permanganate is most likely formed under acidic reaction conditions [61]. Chlorine dioxide is highly explosive and tends to decompose at about 60 °C. Thus, chlorate as oxidant is less favored compared to potassium permanganate. In general, about 2–5 mass equivalents of the oxidant are used and the reaction lasts typically a few hours. Third, the oxidized graphite, GrO, is isolated by addition of water to the sulfuric acid, even if this procedure heats up the reaction mixture to about 90 °C. If potassium permanganate is used as oxidant, hydrogen peroxide will subsequently be necessary to make manganese salts soluble to remove them from the product. Purification involves removal of inorganic salts, either by repeated washing of GrO with water or/and hydrochloric acid. Alternatively, dialysis is used for purification. The GrO is yielded and can be isolated as yellow to brown material. In the fourth step, GO is delaminated from GrO. This delamination is either achieved by shaking or elongated stirring. However, sonication accelerates this process. Some parts of GrO may not delaminate to GO and therefore, the dispersions are centrifuged to remove them (Figure 2.20).

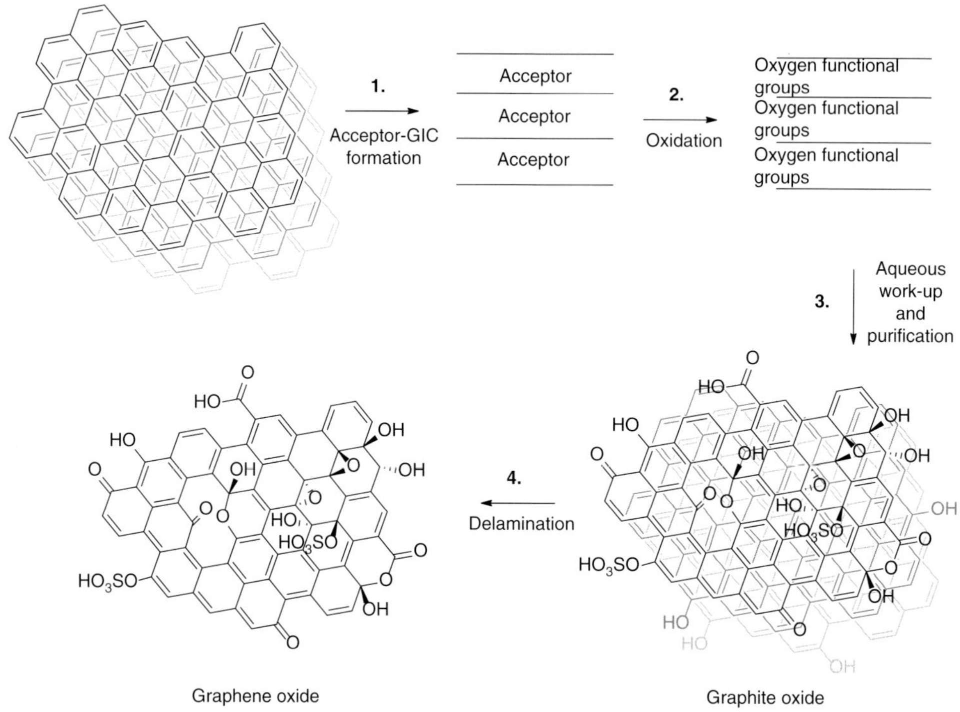

Figure 2.19 General illustration of the synthesis of GO. Graphite becomes intercalated and activated, followed by oxidation. Aqueous work-up yields graphite oxide and GO after delamination. (Reprinted from Ref. [58] with permission from Springer Science+Business Media, Copyright 2014.)

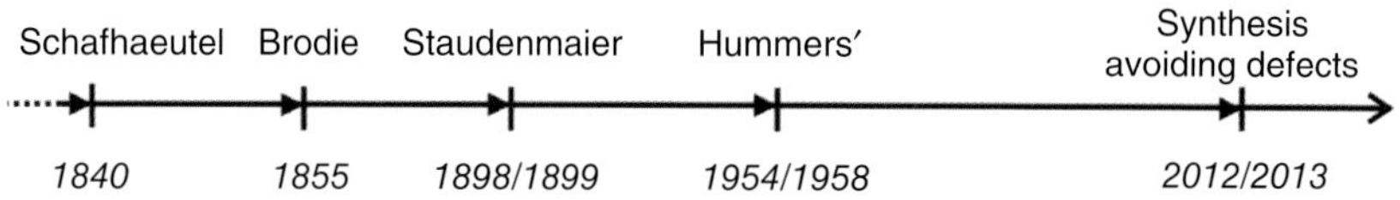

Figure 2.20 Historical development of the synthesis of GrO and GO, respectively. (Reprinted from Ref. [58] with permission from Springer Science+Business Media, Copyright 2014.)

2.7.2.2 Historical Development of the Synthesis of GrO

The synthesis of GrO, also called graphitic acid, has a long history that started in 1840 [57, 61]. The synthetic procedures developed within roughly 60 years and were further enhanced within the last 100 years. Only recently, it was possible to determine the density of defects in the carbon framework and novel methods could be found [60, 71, 72].

Schafhaeutel observed the expansion of graphite in hot sulfuric acid by adding nitric oxide to a dispersion of graphite in 1840 [73]. This oxidation procedure was repeated and finally a clear reaction mixture was obtained. It seems very likely that GrO was formed during these early experiments. However, the first preparation and isolation of GrO was described by Brodie in 1855. The product was termed as graphitic acid. The oxidation of graphite was performed in nitric acid using potassium chlorate as oxidant. The procedure had to be repeated several times to yield a yellow dispersion of GrO. Chlorine dioxide can accumulate and spontaneous decomposition may occur by this procedure. Staudenmaier optimized the method and found a less dangerous and faster two-step protocol. First, the dispersion of graphite is treated in a mixture of sulfuric acid and nitric acid with potassium chlorate as oxidant. Second, the partially oxidized product is further oxidized with potassium permanganate to GrO. In 1909 Charpy published a further improved synthesis of the desired yellow GrO. Graphite is dispersed in sulfuric acid and oxidized by addition of portions of potassium permanganate. It is described that the temperature must be kept below 45 °C not to rapidly over-oxidize the product. Higher temperatures lead to complete decomposition of the reaction mixture. Hummers and Offeman published the scalable synthesis of GrO by a similar method as described by Charpy in the 1950s [57]. This procedure is called Hummers' method and was the most often used protocol for the synthesis of GO within recent years. The yield of the oxidation process generally depends on the type graphite and parameters, such as the flake size and crystallinity of graphite, the reaction time, temperature and amount of oxidant or the purity of the graphite. The successful preparation of GrO and GO, respectively, is generally indicated by the transformation of black graphite to yellow GrO. Furthermore, the yield can be increased by pre-oxidizing graphite with persulfate in sulfuric acid. All methods have in common that GrO is yielded as brown to yellow solid after removing the solvent and after delamination of GrO to individual layers of GO.

Moreover, all the described methods have in common that significant amounts of defects are introduced into the carbon lattice during the synthesis [62]. These defects are permanent defects, due to CO_2 formation. The amount of defects depends on the exact reaction protocol. Raman spectroscopy can be applied on single layers of reduced GO that have been deposited on a solid support, such as a Si-wafer with a grown SiO_2 layer with 300 nm thickness. The reduction is done within about 15 min using a vapor of hydriodic acid and trifluoroacetic acid at 80 °C. The density of defects can subsequently be determined by the analysis of SRM data.

The preparation of GO that bears defect densities below 0.1% can be achieved by oxidizing graphite in sulfuric acid with potassium permanganate and keeping the reaction temperature below 5–10 °C. The work-up process must also be accomplished avoiding local heat higher than 5–10 °C to prevent over-oxidation. GO with an almost intact carbon framework is yielded by this method and the density of defects within the carbon framework is minimized. Thus, the honeycomb lattice is largely intact, as it is in graphene. Therefore, this material can be called oxo-functionalized graphene, which indicates that the graphene lattice is almost preserved and oxygen functional groups are added to the π-surface. Furthermore, functional groups at edges of flakes or defects play only a minor role for the chemistry of this material, in contrast to defective GO. The low defect concentration is important for using GO as a precursor generating graphene.

Electrical properties are dependent on the chemical structure and graphene derived from oxo-functionalized graphene exhibits charge carrier mobility values in the range of 100–1000 $cm^2 V^{-1} s^{-1}$. In contrast, graphene derived from GO, also called reduced GO, exhibits a higher defect density and thus, charge carrier mobility values depend on hopping processes and have been determined to roughly 0.1–10 $cm^2 V^{-1} s^{-1}$ [62].

2.7.2.3 Structure of GO

The structure of GO is highly variable. The absolute structure depends on the preparation conditions. Thus, no molecular equation can be given and no precise structure can be formulated. However, in general, with oxidizing graphite toward GO, oxo-functional groups are introduced. Ideally a hexagonal σ-framework of carbon atoms is the basis. This carbon framework is normally ruptured as outlined above due to over-oxidation. Therefore, the honeycomb lattice is not preserved and σ-hole defects are generated in amounts of about 3%. By controlling the oxidation procedure, the loss of carbon can be minimized and average densities of defects of about 0.01% have been demonstrated [62, 66, 74].

Oxygen functional groups decorate both sides of the basal plane in GO. However, the amount and ratio of groups is arbitrary. About every second carbon atom is functionalized, as in part illustrated in Figure 2.21. It seems very likely that patches of sp^2-hybridized carbon are part of the structure. The growth of the sp^2-patches may increase with time and thus, the initial chemical structure can be interpreted as metastable [75]. However, covalent functional groups are mainly hydroxyl and epoxy groups. If sulfuric acid is used for the preparation of GO, covalently bound organosulfate groups will be generated. About 1 organosulfate on 30 carbon atoms has been reported [62]. Among other functional groups present on the basal plane are functional groups at edges of flakes and defect sites. The nature

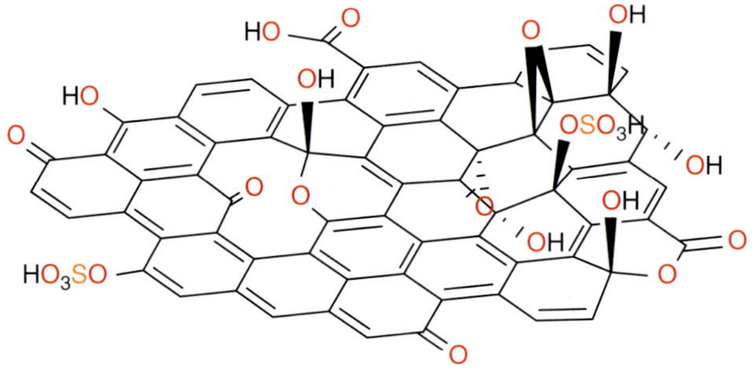

Figure 2.21 Chemical sketch for the illustration of functional groups in GO that bears defects. At both sides of the basal plane, there are hydroxyl groups, epoxy groups, and organosulfate groups. At edges of defects and edges of flakes there are possibly hydroxyl groups, ketones, actetals, carboxylic acids, lactol groups, and organosulfate groups in arbitrary amounts. A defect with a proposed structure that comprises one carbonyl group and a hemi-acetal is shown as well. (From Ref. [2] with permission from Wiley-VCH Verlag GmbH & Co, Copyright 2014.)

of these groups is not precisely known, but carboxylic acids, lactol, carbonyl and hydroxyl groups and also organosulfate groups are very likely bound. At point defects also acetal structures are plausible to be present. These structure motives are in part considered in the models of Lerf and Klinowski, Gao *et al.*, and Eigler *et al.* [2]. The development of these models is mainly based on NMR investigations, thermogravimetric analysis, infrared spectroscopy, and others. Another type of heterogeneity is due to the lateral dimension of flakes that ranges between tens of nanometers to hundreds of micrometers. In addition, high-resolution transmission electron microscopy was used to evaluate the structure of GO and preserved regions of graphene like structures with lateral dimensions of about 1 nm were identified. About 80% of disordered regions are found, which were correlated to functionalized regions [66]. Furthermore, holes with diameters of about 1 nm were also visualized.

2.7.2.4 GO as Precursor for Graphene

Graphene can be prepared from GO by reduction. The quality of the obtained graphene is heterogeneous by means of the density of defects within the carbon framework. Structural defects in graphene are illustrated in Figure 2.23a. Various methods, including thermal or light driven disproportionation and chemical reduction using various reductants, have been reported [76]. The reducing agent may even be incorporated in the carbon lattice, for example for hydrazine. In addition, metal hydrides, hydriodic acid, sulfur containing reducing agents, alcohols, ascorbic acid (AS), sugars, metals in acids, amino acids, plant extracts, and even microorganism were used. Moreover, it was demonstrated that defects in GO limit the final quality of graphene as well as the type of reducing agent [62].

Thermal disproportionation of GO is accompanied by a mass loss of about 40% up to 300 °C and CO_2, CO and H_2O are mainly formed [62]. The success of the chemical reduction of GO is often evaluated by the *C*/*O* ratio according to X-ray photoelectron spectroscopy (XPS). However, a quantitative evaluation is difficult due to impurities and structural defects. Especially for low concentrations of defects results obtained by XPS analysis may be disturbed by impurities. Furthermore, the *C*/*O* ratio cannot account for structural defects such as five-membered rings beneath seven-membered rings. Such structures can also not be termed as graphene and they reflect a defect that is detected by Raman spectroscopy. Persistent oxygen functional groups may remain at defect sites after reduction; however, quantification is barely possible.

The most reliable evaluation of the quality of graphene produced from GO is by using Raman spectroscopy. SRM is the method of choice for a reliable quantification of the quality [62]. Investigations indicate that the density of defects can be as low as 0.01% or up to several percent (Figure 2.23b).

The chemical reduction of GO placed on a substrate using a potent reduction agent can be highly efficient and termed as quantitative. As illustrated in Figure 2.22, hydriodic acid provides both electrons and protons. Iodide is oxidized and iodine is formed. The electrons can easily reduce GO; that is, oxo-functional groups, such as epoxy groups or hydroxyl groups are protonated.

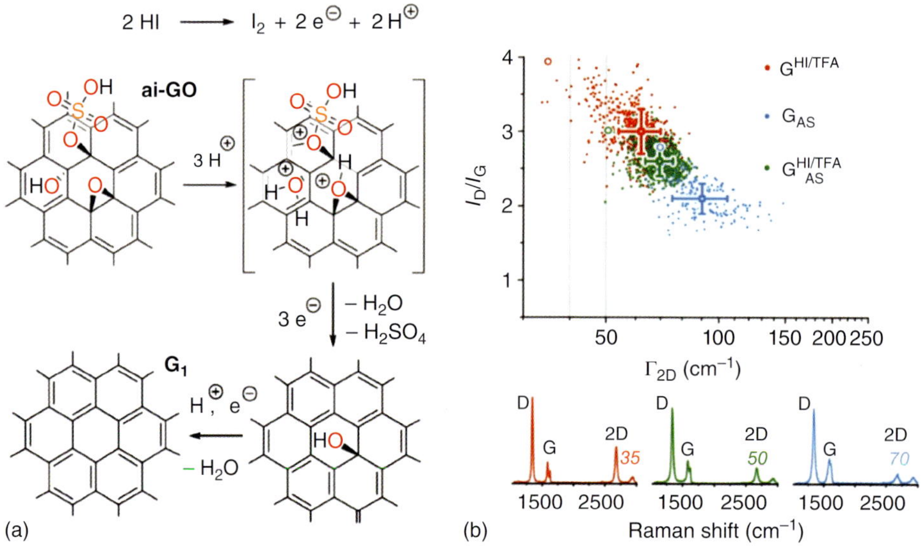

Figure 2.22 (a) Proposed general reduction mechanism of GO with an almost intact carbon framework, also called oxo-functionalized graphene. Oxo-groups attached to graphene are protonated and reductively cleaved on both sides of the basal plane. (b) Raman spectroscopic studies reveal that both sides of the basal plane are efficiently reduced, even if only one side of oxo-functionalized graphene is accessible for the reducing agent. Red, Reduction with HI/TFA from the top; blue, reduction with ascorbic acid (AS) from the lower side; and green, stepwise reduction from both sides [77].

Water molecules and organosulfate sulfuric acid are formed and cleaved as good leaving groups. Finally, graphene is yielded. Even if the reduction agent can access only from the top and not from the bottom, because the flakes are placed on a substrate, the reduction proceeds at least as sufficient enough for the reducing agent to access both sides of the basal plane. As proof, AS was added to the subphase in the Langmuir–Blodgett trough to give the reducing agent access to the lower side of GO. Subsequently, the reduced flakes were transferred to a substrate and additionally reduced by HI/TFA: hydriodic acid / trifluoroacedic acid (HI/TFA) vapor that can access only from the top. SRM data reveal that HI/TFA reduction alone is highly sufficient, as depicted in Figure 2.22b.

The wet-chemical synthesis of graphene is possible from GO with defects on the %-scale, almost intact GO with an average density of defects of about 0.03%, or oxo-functionalized graphene with an average density of defects of 0.05%. The processes for the controlled oxidation of graphite are illustrated in Figure 2.23b and c. Avoiding local heat during oxidation of graphite in sulfuric acid with potassium permanganate is crucial, as is the control of temperature during aqueous work-up. The individualized and almost intact GO can be placed on a surface for reduction to graphene (Figure 2.23c).

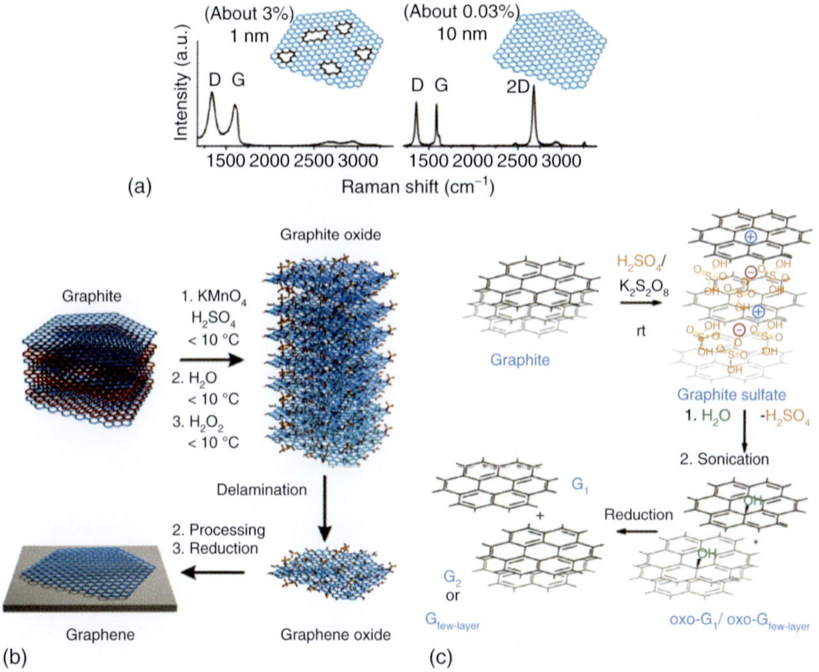

Figure 2.23 (a) Illustration of the defect density in graphene derived from GO with different defect densities; left: σ-hole defects within the carbon framework with a defect density of several percentage values (residual functional groups omitted); right: graphene from oxo-functionalized graphene with an almost intact carbon framework and Raman spectra of derived graphene. (b) Schematic of the synthesis of oxo-functionalized graphene with an almost intact carbon framework and related graphene. (From Ref. [2] with permission from Wiley-VCH Verlag GmbH & Co, Copyright 2014.) (c) Schematic of the synthesis of oxo-functionalized graphene from graphite sulfate and related graphene. (From Ref. [72] with permission from Wiley-VCH Verlag GmbH & Co, Copyright 2015.)

An alternative graphene synthesis from oxo-functionalized graphene with a degree of functionalization of about 4% is also possible. The graphite is stirred in sulfuric acid with an oxidant, such as potassium persulfate, to form graphite sulfate. In graphite sulfate the layers of graphene are positively charged and during aqueous work-up hydroxyl groups are mainly added. The product can in part be delaminated to oxo-functionalized graphene and few-layer graphene. After placement on a substrate, graphene can be prepared by reduction (Figure 2.23c).

The quality of graphene derived from almost intact GO was determined by SRM and the data are shown in Figure 2.24a. The FWHM of the 2D peak is about 50 cm^{-1} for single layers, which relates to a density of defects of about 3%. The statistical analysis by Raman spectroscopy of films of flakes of graphene derived

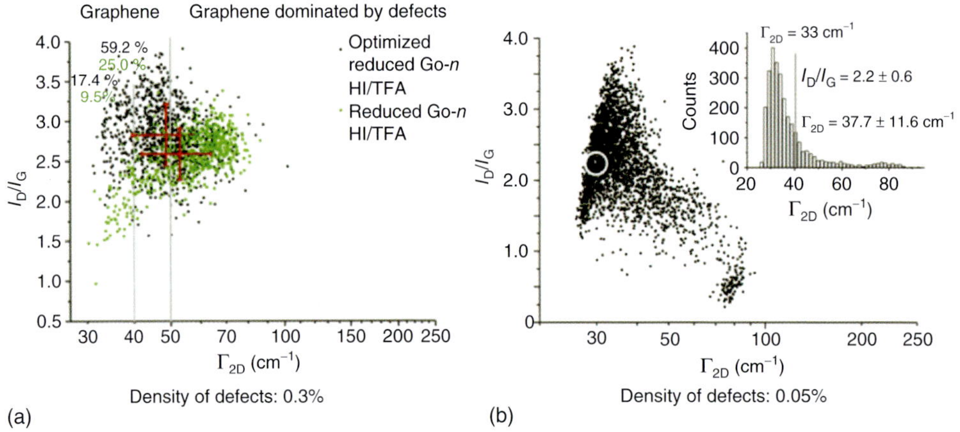

Figure 2.24 SRM data for graphene from (a) almost intact GO prepared by the method illustrated in Figure 2.23c, data relate to a density of defects of about 0.3% [12]. (b) Oxo-functionalized graphene derived from graphite sulfate, as illustrated in Figure 2.23d and the average density of defects can be related to about 0.05% [72].

from oxo-functionalized graphene with a degree of functionalization of about 4% was also determined, and the FWHM of the 2D peak was determined to be about 33 cm^{-1} and the I_D/I_G ratio to about 2 in average (Figure 2.24b). These data suggest that the density of defects is roughly only 0.05%. Especially, a value of 33 cm^{-1} for the 2D peak can be related to graphene.

Although progress has been made in recent years, methods must be further optimized to increase the yield and purity of graphene. Moreover, the density of defects should also be further reduced not to show any D peak in Raman spectra. Nevertheless, even if the GO-synthesis that provides a variable density of defects is very young, one can assume that the performance of graphene applications will increase using graphene with controlled amounts of defects.

2.8 Synthesis of Nanographene from Small Molecules

It was demonstrated that graphene can be prepared either wet-chemically at high temperature from small molecules or directly from graphite. However, as discussed earlier, the yield of graphene among few-layer graphene and the quality are statistically distributed. Precise control of the size of flakes is difficult, even if separation techniques improve [2, 78]. While the growth of graphene crystals on metal surfaces is improving, precise control of size and especially shape remains an unsolved problem [79]. The outstanding physical properties of graphene are due to the chemical structure of the honeycomb lattice. For example, graphene is a zero band-gap semiconductor [34]. A band-gap can be introduced by narrowing the width of a sheet of graphene down to few nanometers (Figure 2.25) [82].

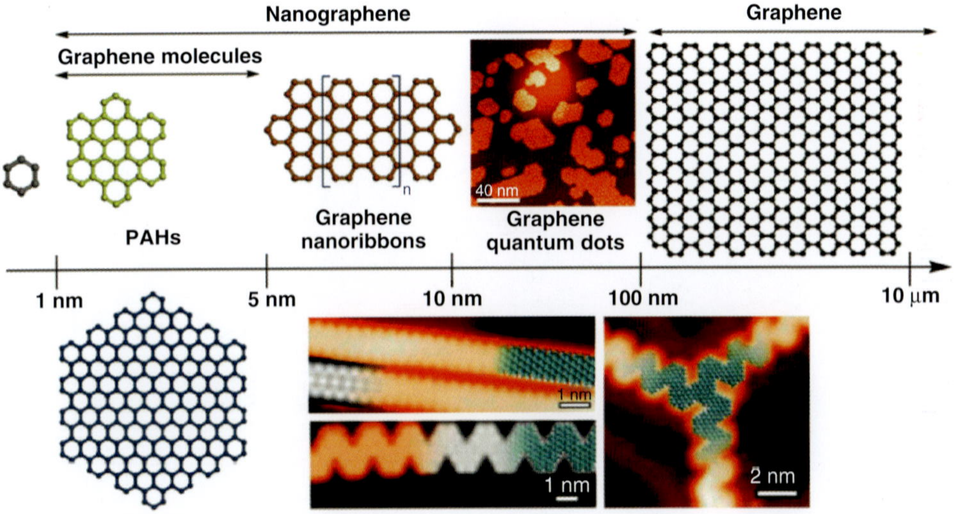

Figure 2.25 Illustration of graphene derived structures ranging from molecules, such as benzene or hexabenzocorronene and even larger units, such as C_{222} graphene like units. Structures are monodisperse with a diameter of up to 5 nm. Graphene nanoribbons of larger size and variable shape; graphene quantum dots (GQDs) with lateral dimensions between 10 and 100 nm and graphene of typically micrometer size crystals, at least >100 nm in both directions. (From Ref. [80] with permission from Wiley-VCH Verlag GmbH & Co, Copyright 2012.) The STM images of GQDs modified with permission from Ref. [81]. (Reprinted from Ref. [82] with permission from Nature Publishing Group, Copyright 2010.)

Such structures exhibit a high aspect ratio by means of length and width, keeping the thickness at the atomic level. It was found that the band-gap relates to the width of the graphene nanoribbons [83, 84]. Thus, graphene nanoribbons are very interesting candidates for future applications. However, the precise synthesis of graphene of various shapes and sizes that can range from few nanometers to micrometer size remains elusive. In addition, control over the configuration of edges of graphene that can be either zigzag or armchair is highly desired but is still a challenging task. Therefore, chemists facilitate the so-called bottom-up approach to gain control over these challenges [85].

Precise chemical reactions are used to build defined chemical structures. Examples of such carbon-rich molecules are depicted in Figure 2.25. The size of synthetically accessible molecules is on the nanometer-scale. The synthetic approach for a molecule whose diameter approaches 5 nm is shown in Figure 2.26. The graphene-like core unit is C_{222} [86]. The precursor molecule is still soluble in organic solvents due to its three dimensional structure. Such large molecules can be synthesized by three steps only. The first step involves a Diels–Alder reaction of the triyne and two equivalents of tetraphenylcyclopentadienone at 190 °C using diphenyl ether as solvent. Only the outer acetylene units are reactive

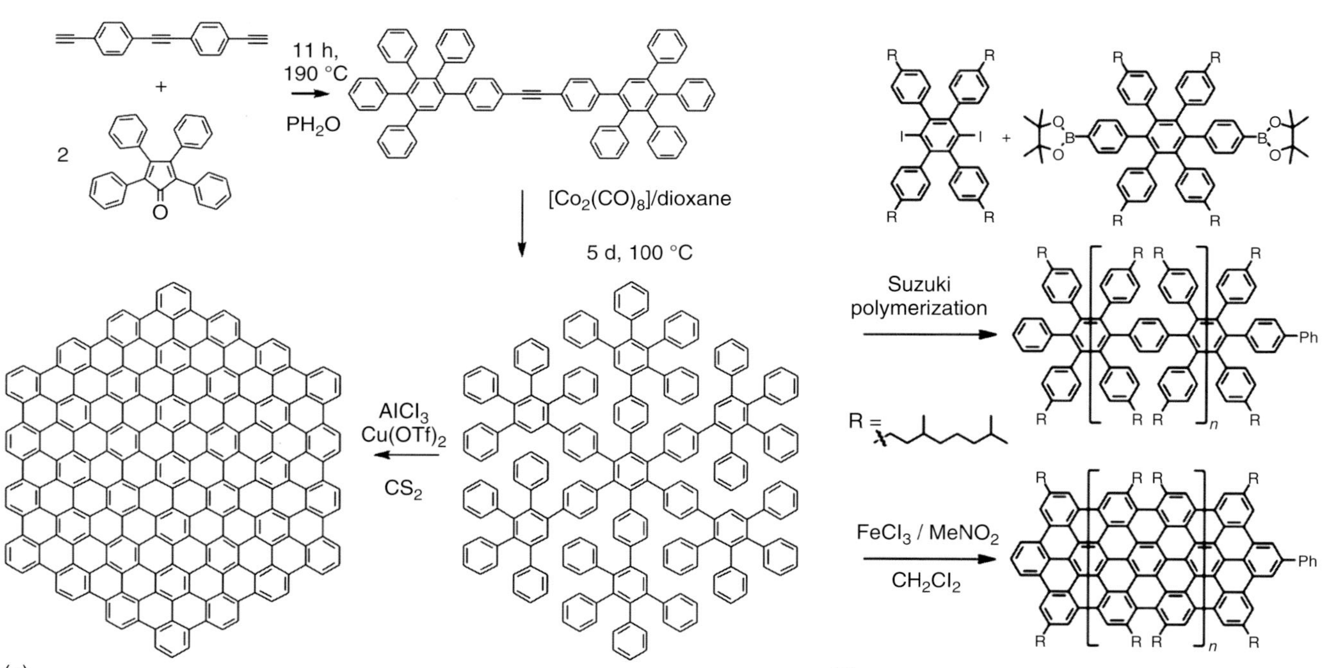

(a) (b)

Figure 2.26 (a) Example of the synthesis of the C_{222} graphene-like unit by the wet-chemical bottom-up approach. The reaction scheme shows a Diels–Alder reaction followed by cyclotrimerization that yields the soluble precursor molecule that can be oxidized in CS_2 to the depicted π-conjugated hydrocarbon [86]. (b) Synthetic bottom-up approach to graphene nanoribbons. Soluble precursors functionalized by halides and boronic esters are C–C coupled to define the nanoribbon structure. Oxidation of the preoriented benzene rings yields π-conjugated nanoribbons. (Modified from Ref. [87] with permission from the American Chemical Society, Copyright 2008.)

under these reaction conditions. The residual triple bond is still reactive toward cyclotrimerization and, therefore, yields the three dimensional soluble precursor molecule with 37 benzene units. Treating this precursor under oxidative conditions with a mixture of aluminum chloride and copper triflate in carbon disulfide affords a black solid that can be identified as the target molecule with a C_{222} graphene-like unit by mass spectrometry.

Chemical strategies are a very promising way toward the controlled synthesis of graphene nanostructures, such as graphene nanoribbons [80, 87]. Graphene nanoribbons are synthetically accessible by the bottom-up approach. Small and defined precursor molecules are synthesized to enable the Suzuki reaction (Figure 2.26b). Therefore, tetraphenyl-substituted and iodine-activated benzene is coupled with a hexabenzene derivative bearing boronic esters. These soluble precursors units define the graphene nanoribbon by repeated coupling to polymers. Finally, the polymer is cross-coupled by oxidation using Scholl-reaction conditions to yield the π-conjugated nanoribbon.

The bottom-up approach is extremely versatile due to the link to organic chemistry. A set of differently shaped nanoribbons can be prepared, such as the chevron-like nanoribbons, by using defined precursor molecules [82]. As depicted in Figure 2.27a, a chevron structure of nanoribbons is yielded using the suitable brominated precursor molecule. Here, a hexaphenyl benzene derivative is used, which bears a brominated biphenyl unit. The linear coupling reaction of this precursor defines the final length of the nanoribbon. The oxidation of that polymer leads to the π-conjugated nanoribbon. Unfortunately, the molecular weight of such nanoribbons does no longer allow a standard characterization, such as nuclear magnetic resonance spectroscopy, as usually used in organic chemistry. However, the nanoribbons can be visualized with atomic precision by scanning probe microscopy. In another example, a precursor molecule with two coupling groups and three coupling groups are reacted and as predefined by the chemical structure, a junction is obtained as shown in Figure 2.27b [82]. Other graphene-like structures, such as porous graphene, can also be synthesized. Cyclohexaphane derivatives can be used for such a polymerization approach. Finally, a porous 2D-polymer is obtained (Figure 2.27) [88].

Until now, there are manifold concepts for the bottom-up approach. However, only selected examples are discussed here to illustrate the basic concepts and strategies of the current bottom-up approach. The greatest benefit is the atomic precision of chemical structures and super-structures formed after polymerization. In some examples, metal surfaces are necessary to generate the final π-conjugated system. Moreover, the final molecules are not soluble and not transferable from one substrate to another. Even if they could be transferred, defects must be considered. Another hurdle is due to the poor control of size, especially for polymerized structures, such as graphene nanoribbons. Nevertheless, the prospects of the bottom-up approach are great and will certainly find applications in the future.

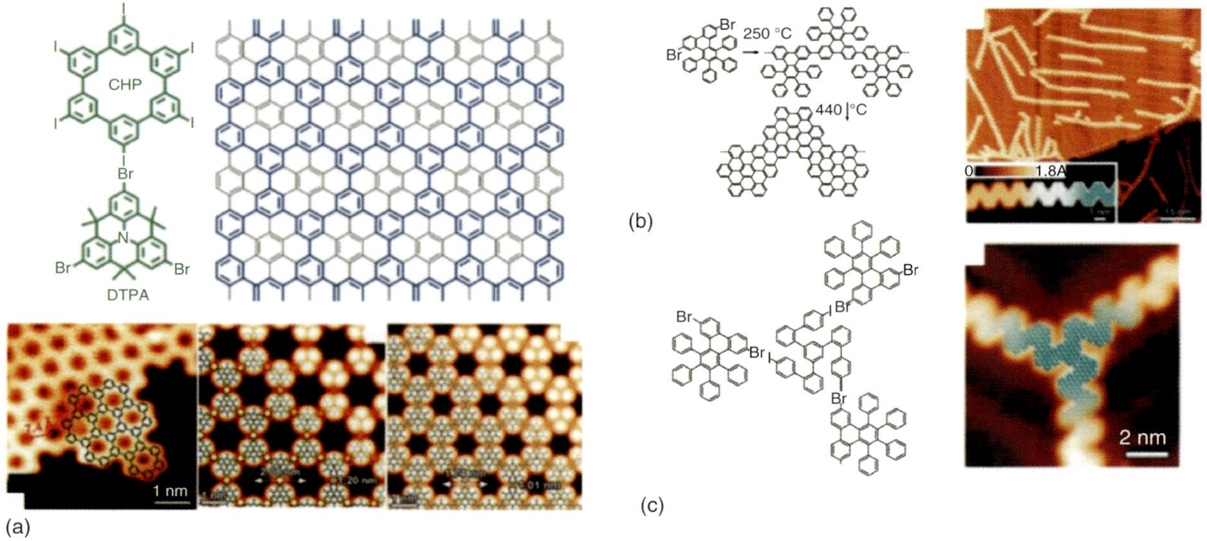

Figure 2.27 (a) Top: Chemical structure of a cyclohexaphane derivative as precursor for porous graphene (blue, illustrated on gray graphene structure). Bottom: Related high-resolution scanning probe microscope image. (From Ref. [80] with permission from Wiley-VCH Verlag GmbH & Co, Copyright 2012; modified from [88] with permission from the Royal Society of Chemistry. Copyright 2008; images reproduced from Ref. [89] with permission from the Royal Society of Chemistry, Copyright 2011).) (b) Left: Synthesis of chevron-type graphene nanoribbons by the polymerization of bis-brominated tetraphenyl triphenylene at 250 °C to a prepolymer followed by dehydrogenation at 440 °C. Right: Scanning probe microscope image visualizing the chevron-type structure. (c) Bottom-up synthesis of a junction using bis-activated and tris-activated precursor molecules and scanning probe microscopy image of the junction. (Reprinted from Ref. [82] with permission from the Nature Publishing Group, Copyright 2010.)

References

1. Fitzer, E., Kochling, K.H., Boehm, H.P., and Marsh, H. (1995) Recommended terminology for the description of carbon as a solid (IUPAC Recommendations 1995). *Pure Appl. Chem.*, **67** (3), 473–506.
2. Eigler, S. and Hirsch, A. (2014) Chemistry with graphene and graphene oxide-challenges for synthetic chemists. *Angew. Chem. Int. Ed.*, **53** (30), 7720–7738.
3. He, K., Lee, G.D., Robertson, A.W., Yoon, E., and Warner, J.H. (2014) Hydrogen-free graphene edges. *Nat. Commun.*, **5**, 3040.
4. MacDonald, A.H. and Bistritzer, R. (2011) Materials science: graphene moire mystery solved? *Nature*, **474** (7352), 453–454.
5. Koehler, F.M. and Stark, W.J. (2013) Organic synthesis on graphene. *Acc. Chem. Res.*, **46** (10), 2297–2306.
6. Ferrari, A.C. and Basko, D.M. (2013) Raman spectroscopy as a versatile tool for studying the properties of graphene. *Nat. Nanotechnol.*, **8** (4), 235–246.
7. Englert, J.M., Vecera, P., Knirsch, K.C., Schafer, R.A., Hauke, F., and Hirsch, A. (2013) Scanning-Raman-microscopy for the statistical analysis of covalently functionalized graphene. *ACS Nano*, **7** (6), 5472–5482.
8. Eigler, S., Hof, F., Enzelberger-Heim, M., Grimm, S., Müller, P., and Hirsch, A. (2014) Statistical-Raman-microscopy and atomic-force-microscopy on heterogeneous graphene obtained after reduction of graphene oxide. *J. Phys. Chem. C*, **118** (14), 7698–7704.
9. Cançado, L.G., Jorio, A., Ferreira, E.H.M., Stavale, F., Achete, C.A., Capaz, R.B., Moutinho, M.V.O., Lombardo, A., Kulmala, T.S., and Ferrari, A.C. (2011) Quantifying defects in graphene via Raman spectroscopy at different excitation energies. *Nano Lett.*, **11** (8), 3190–3196.
10. Lucchese, M.M., Stavale, F., Ferreira, E.H.M., Vilani, C., Moutinho, M.V.O., Capaz, R.B., Achete, C.A., and Jorio, A. (2010) Quantifying ion-induced defects and Raman relaxation length in graphene. *Carbon*, **48** (5), 1592–1597.
11. Blake, P., Hill, E.W., Castro Neto, A.H., Novoselov, K.S., Jiang, D., Yang, R., Booth, T.J., and Geim, A.K. (2007) Making graphene visible. *Appl. Phys. Lett.*, **91** (6), 063124.
12. Eigler, S., Grimm, S., Enzelberger-Heim, M., Müller, P., and Hirsch, A. (2013) Graphene oxide: efficiency of reducing agents. *Chem. Commun.*, **49** (67), 7391–7393.
13. Englert, J.M., Dotzer, C., Yang, G., Schmid, M., Papp, C., Gottfried, J.M., Steinrück, H.-P., Spiecker, E., Hauke, F., and Hirsch, A. (2011) Covalent bulk functionalization of graphene. *Nat. Chem.*, **3** (4), 279–286.
14. Badami, D.V. (1962) Graphitization of α-silicon carbide. *Nature*, **193** (4815), 569–570.
15. Virojanadara, C., Syvajarvi, M., Yakimova, R., Johansson, L.I., Zakharov, A.A., and Balasubramanian, T. (2008) Homogeneous large-area graphene layer growth on 6H-SiC(0001). *Phys. Rev. B*, **78** (24), 245403.
16. Emtsev, K.V., Speck, F., Seyller, T., Ley, L., and Riley, J.D. (2008) Interaction, growth, and ordering of epitaxial graphene on SiC{0001} surfaces: a comparative photoelectron spectroscopy study. *Phys. Rev. B*, **77** (15), 155303.
17. Borovikov, V. and Zangwill, A. (2009) Step-edge instability during epitaxial growth of graphene from SiC(0001). *Phys. Rev. B*, **80** (12), 121406R.
18. Emtsev, K.V., Bostwick, A., Horn, K., Jobst, J., Kellogg, G.L., Ley, L., McChesney, J.L., Ohta, T., Reshanov, S.A., Rohrl, J., Rotenberg, E., Schmid, A.K., Waldmann, D., Weber, H.B., and Seyller, T. (2009) Towards wafer-size graphene layers by atmospheric pressure graphitization of silicon carbide. *Nat. Mater.*, **8** (3), 203–207.
19. Riedl, C., Coletti, C., Iwasaki, T., Zakharov, A.A., and Starke, U. (2009) Quasi-free-standing epitaxial graphene on SiC obtained by hydrogen intercalation. *Phys. Rev. Lett.*, **103** (24), 246804.
20. Sutter, P. (2009) Epitaxial graphene: how silicon leaves the scene. *Nat. Mater.*, **8** (3), 171–172.

21. Kazakova, O., Panchal, V., and Burnett, T. (2013) Epitaxial graphene and graphene-based devices studied by electrical scanning probe microscopy. *Crystals*, **3** (1), 191–233.
22. Gorantla, S., Bachmatiuk, A., Hwang, J., Alsalman, H.A., Kwak, J.Y., Seyller, T., Eckert, J., Spencer, M.G., and Rummeli, M.H. (2014) A universal transfer route for graphene. *Nanoscale*, **6** (2), 889–896.
23. Riedl, C., Coletti, C., and Starke, U. (2010) Structural and electronic properties of epitaxial graphene on SiC(0 0 0 1): a review of growth, characterization, transfer doping and hydrogen intercalation. *J. Phys. D: Appl. Phys.*, **43** (37), 374009.
24. Avouris, P. and Dimitrakopoulos, C. (2012) Graphene: synthesis and applications. *Mater. Today*, **15** (3), 86–97.
25. Mattevi, C., Kima, H., and Chhowalla, M. (2011) A review of chemical vapour deposition of graphene on copper. *J. Mater. Chem.*, **21** (10), 3324–3334.
26. Zhang, Y., Zhang, L., and Zhou, C. (2013) Review of chemical vapor deposition of graphene and related applications. *Acc. Chem. Res.*, **46** (10), 2329–2339.
27. Yu, Q., Lian, J., Siriponglert, S., Li, H., Chen, Y.P., and Pei, S.-S. (2008) Graphene segregated on Ni surfaces and transferred to insulators. *Appl. Phys. Lett.*, **93** (11), 113103.
28. Zhang, Y., Gomez, L., Ishikawa, F.N., Madaria, A., Ryu, K., Wang, C., Badmaev, A., and Zhou, C. (2010) Comparison of graphene growth on single-crystalline and polycrystalline Ni by chemical vapor deposition. *J. Phys. Chem. Lett.*, **1** (20), 3101–3107.
29. Blume, R., Kidambi, P.R., Bayer, B.C., Weatherup, R.S., Wang, Z.J., Weinberg, G., Willinger, M.G., Greiner, M., Hofmann, S., Knop-Gericke, A., and Schlogl, R. (2014) The influence of intercalated oxygen on the properties of graphene on polycrystalline Cu under various environmental conditions. *Phys. Chem. Chem. Phys.*, **16** (47), 25989–26003.
30. Hao, Y., Bharathi, M.S., Wang, L., Liu, Y., Chen, H., Nie, S., Wang, X., Chou, H., Tan, C., Fallahazad, B., Ramanarayan, H., Magnuson, C.W., Tutuc, E., Yakobson, B.I., McCarty, K.F., Zhang, Y.W., Kim, P., Hone, J., Colombo, L., and Ruoff, R.S. (2013) The role of surface oxygen in the growth of large single-crystal graphene on copper. *Science*, **342** (6159), 720–723.
31. Lee, J.H., Lee, E.K., Joo, W.J., Jang, Y., Kim, B.S., Lim, J.Y., Choi, S.H., Ahn, S.J., Ahn, J.R., Park, M.H., Yang, C.W., Choi, B.L., Hwang, S.W., and Whang, D. (2014) Wafer-scale growth of single-crystal monolayer graphene on reusable hydrogen-terminated germanium. *Science*, **344** (6181), 286–289.
32. Bae, S., Kim, H., Lee, Y., Xu, X., Park, J.-S., Zheng, Y., Balakrishnan, J., Lei, T., Kim, H.R., Song, Y.I., Kim, Y.-J., Kim, K.S., Özyilmaz, B., Ahn, J.-H., Hong, B.H., and Iijima, S. (2010) Roll-to-roll production of 30-inch graphene films for transparent electrodes. *Nat. Nanotechnol.*, **5** (8), 574–578.
33. Kamalakar, M.V., Groenveld, C., Dankert, A., and Dash, S.P. (2015) Long distance spin communication in chemical vapour deposited graphene. *Nat. Commun.*, **6**, 6766.
34. Novoselov, K.S., Geim, A.K., Morozov, S.V., Jiang, D., Zhang, Y., Bubonos, S.V., Grigorieva, I.V., and Firsov, A.A. (2004) Electric field effect in atomically thin carbon films. *Science*, **306** (5696), 666–669.
35. Yi, M. and Shen, Z. (2015) A review on mechanical exfoliation for the scalable production of graphene. *J. Mater. Chem. A*, **3** (22), 11700–11715.
36. Nilsson, J., Castro Neto, A.H., Guinea, F., and Peres, N.M.R. (2008) Electronic properties of bilayer and multilayer graphene. *Phys. Rev. B*, **78** (4), 045405.
37. Novoselov, K.S., McCann, E., Morozov, S.V., Fal'ko, V.I., Katsnelson, M.I., Zeitler, U., Jiang, D., Schedin, F., and Geim, A.K. (2006) Unconventional quantum Hall effect and Berry's phase of 2π in bilayer graphene. *Nat. Phys.*, **2** (3), 177–180.
38. Nacken, T.J., Damm, C., Xing, H., Rüger, A., and Peukert, W. (2015) Determination of quantitative structure–property and structure–process relationships for

graphene production in water. *Nano Res.*, **5** (71), 57328.
39. Ciesielski, A. and Samori, P. (2014) Graphene via sonication assisted liquid-phase exfoliation. *Chem. Soc. Rev.*, **43** (1), 381–398.
40. Damm, C., Nacken, T.J., and Peukert, W. (2015) Quantitative evaluation of delamination of graphite by wet media milling. *Carbon*, **81** (1), 284–294.
41. Khan, U., Porwal, H., O'Neill, A., Nawaz, K., May, P., and Coleman, J.N. (2011) Solvent-exfoliated graphene at extremely high concentration. *Langmuir*, **27** (15), 9077–9082.
42. Green, A.A. and Hersam, M.C. (2009) Solution phase production of graphene with controlled thickness via density differentiation. *Nano Lett.*, **9** (12), 4031–4036.
43. Green, A.A. and Hersam, M.C. (2010) Emerging methods for producing monodisperse graphene dispersions. *J. Phys. Chem. Lett.*, **1** (2), 544–549.
44. Paton, K.R., Varrla, E., Backes, C., Smith, R.J., Khan, U., O'Neill, A., Boland, C., Lotya, M., Istrate, O.M., King, P., Higgins, T., Barwich, S., May, P., Puczkarski, P., Ahmed, I., Moebius, M., Pettersson, H., Long, E., Coelho, J., O'Brien, S.E., McGuire, E.K., Sanchez, B.M., Duesberg, G.S., McEvoy, N., Pennycook, T.J., Downing, C., Crossley, A., Nicolosi, V., and Coleman, J.N. (2014) Scalable production of large quantities of defect-free few-layer graphene by shear exfoliation in liquids. *Nat. Mater.*, **13** (6), 624–630.
45. Englert, J.M., Röhrl, J., Schmidt, C.D., Graupner, R., Hundhausen, M., Hauke, F., and Hirsch, A. (2009) Soluble graphene: generation of aqueous graphene solutions aided by a perylenebisimide-based bolaamphiphile. *Adv. Mater.*, **21** (42), 4265–4269.
46. Kozhemyakina, N.V., Englert, J.M., Yang, G., Spiecker, E., Schmidt, C.D., Hauke, F., and Hirsch, A. (2010) Non-covalent chemistry of graphene: electronic communication with dendronized perylene bisimides. *Adv. Mater.*, **22** (48), 5483–5487.
47. Parvez, K., Li, R., Puniredd, S.R., Hernandez, Y., Hinkel, F., Wang, S., Feng, X., and Müllen, K. (2013) Electrochemically exfoliated graphene as solution-processable, highly conductive electrodes for organic electronics. *ACS Nano*, **7** (4), 3598–3606.
48. Dresselhaus, M.S. and Dresselhaus, G. (2002) Intercalation compounds of graphite. *Adv. Phys.*, **51** (1), 1–186.
49. Enoki, T., Suzuki, M., and Endo, M. (2003) *Graphite Intercalation Compounds and Applications*, Oxford University Press, Oxford.
50. Nalimova, V.A., Guérard, D., Lelaurain, M., and Fateev, O.V. (1995) X-ray investigation of highly saturated Li-graphite intercalation compound. *Carbon*, **33** (2), 177–181.
51. Vallés, C., Drummond, C., Saadaoui, H., Furtado, C.A., He, M., Roubeau, O., Ortolani, L., Monthioux, M., and Pénicaud, A. (2008) Solutions of negatively charged graphene sheets and ribbons. *J. Am. Chem. Soc.*, **130** (47), 15802–15804.
52. Vecera, P., Edelthalhammer, K., Hauke, F., and Hirsch, A. (2014) Reductive arylation of graphene: insights into a reversible carbon allotrope functionalization reaction. *Phys. Status Solidi B*, **251** (12), 2536–2540.
53. Englert, J.M., Knirsch, K.C., Dotzer, C., Butz, B., Hauke, F., Spiecker, E., and Hirsch, A. (2012) Functionalization of graphene by electrophilic alkylation of reduced graphite. *Chem. Commun.*, **48** (41), 5025–5027.
54. Schäfer, R.A., Englert, J.M., Wehrfritz, P., Bauer, W., Hauke, F., Seyller, T., and Hirsch, A. (2013) On the way to graphane-pronounced fluorescence of polyhydrogenated graphene. *Angew. Chem. Int. Ed.*, **52** (2), 754–757.
55. Pumera, M. and Wong, C.H. (2013) Graphane and hydrogenated graphene. *Chem. Soc. Rev.*, **42** (14), 5987–95.
56. Ryu, S., Han, M.Y., Maultzsch, J., Heinz, T.F., Kim, P., Steigerwald, M.L., and Brus, L.E. (2008) Reversible basal plane hydrogenation of graphene. *Nano Lett.*, **8** (12), 4597–4602.
57. Eigler, S. (2011) Transparent and electrically conductive films from chemically derived graphene, in *Physics and Applications of Graphene—Experiments*

(ed S. Mikhailov), InTech, New York, pp. 109–130.
58. Eigler, S. (2014) Graphene oxide, in *Encyclopedia of Polymeric Nanomaterials* (eds S. Kobayashi and K. Müllen), Springer, Berlin, Heidelberg. doi: 10.1007/978-3-642-36199-9_333-1
59. Dimiev, A., Kosynkin, D.V., Alemany, L.B., Chaguine, P., and Tour, J.M. (2012) Pristine graphite oxide. *J. Am. Chem. Soc.*, **134** (5), 2815–2822.
60. Eigler, S., Enzelberger-Heim, M., Grimm, S., Hofmann, P., Kroener, W., Geworski, A., Dotzer, C., Rockert, M., Xiao, J., Papp, C., Lytken, O., Steinrück, H.P., Müller, P., and Hirsch, A. (2013) Wet chemical synthesis of graphene. *Adv. Mater.*, **25** (26), 3583–3587.
61. Dreyer, D.R., Park, S., Bielawski, C.W., and Ruoff, R.S. (2010) The chemistry of graphene oxide. *Chem. Soc. Rev.*, **39** (1), 228–240.
62. Eigler, S. and Hirsch, A. (2014) Chemistry with graphene and graphene oxide—challenges for synthetic chemists. *Angew. Chem. Int. Ed.*, **53**, 7720–7738.
63. Novoselov, K.S., Fal'ko, V.I., Colombo, L., Gellert, P.R., Schwab, M.G., and Kim, K. (2012) A roadmap for graphene. *Nature*, **490** (7419), 192–200.
64. Chang, H. and Wu, H. (2013) Graphene-based nanomaterials: synthesis, properties, and optical and optoelectronic applications. *Adv. Funct. Mater.*, **23** (16), 1984–1997.
65. Dai, L. (2013) Functionalization of graphene for efficient energy conversion and storage. *Acc. Chem. Res.*, **46** (1), 31–42.
66. Chen, D., Feng, H., and Li, J. (2012) Graphene oxide: preparation, functionalization, and electrochemical applications. *Chem. Rev.*, **112** (11), 6027–6053.
67. Sun, X., Sun, H., Li, H., and Peng, H. (2013) Developing polymer composite materials: carbon nanotubes or graphene? *Adv. Mater.*, **25** (37), 5153–5176.
68. Lü, K., Zhao, G., and Wang, X. (2012) A brief review of graphene-based material synthesis and its application in environmental pollution management. *Chin. Sci. Bull.*, **57** (11), 1223–1234.
69. Chung, C., Kim, Y.K., Shin, D., Ryoo, S.R., Hong, B.H., and Min, D.H. (2013) Biomedical applications of graphene and graphene oxide. *Acc. Chem. Res.*, **46** (10), 2211–2224.
70. Feng, L., Wu, L., and Qu, X. (2013) New horizons for diagnostics and therapeutic applications of graphene and graphene oxide. *Adv. Mater.*, **25** (2), 168–186.
71. Eigler, S., Dotzer, C., and Hirsch, A. (2012) Visualization of defect densities in reduced graphene oxide. *Carbon*, **50** (10), 3666–3673.
72. Eigler, S. (2015) Graphite sulphate—a precursor to graphene. *Chem. Commun.*, **51** (15), 3162–3165.
73. Schafhaeutl, C. (1840) Ueber die Verbindungen des Kohlenstoffes mit Silicium, Eisen und andern Metallen, welche die verschiedenen Gattungen von Gusseisen, Stahl und Schmiedeeisen bilden. *J. Prakt. Chem.*, **21** (1), 129–157.
74. Georgakilas, V., Otyepka, M., Bourlinos, A.B., Chandra, V., Kim, N., Kemp, K.C., Hobza, P., Zboril, R., and Kim, K.S. (2012) Functionalization of graphene: covalent and non-covalent approaches, derivatives and applications. *Chem. Rev.*, **112** (11), 6156–6214.
75. Kim, S., Zhou, S., Hu, Y., Acik, M., Chabal, Y.J., Berger, C., de Heer, W., Bongiorno, A., and Riedo, E. (2012) Room-temperature metastability of multilayer graphene oxide films. *Nat. Mater.*, **11** (6), 544–549.
76. Chua, C.K. and Pumera, M. (2014) Chemical reduction of graphene oxide: a synthetic chemistry viewpoint. *Chem. Soc. Rev.*, **43** (1), 291–312.
77. Eigler, S. (2014) Mechanistic insights into the reduction of graphene oxide addressing its surfaces. *Phys. Chem. Chem. Phys.*, **16** (37), 19832–19835.
78. Walter, J., Nacken, T.J., Damm, C., Thajudeen, T., Eigler, S., and Peukert, W. (2015) Determination of the lateral dimension of graphene oxide nanosheets using analytical ultracentrifugation. *Small*, **11** (7), 814–825.
79. Luo, Z., Kim, S., Kawamoto, N., Rappe, A.M., and Johnson, A.T. (2011) Growth mechanism of hexagonal-shape graphene flakes with zigzag edges. *ACS Nano*, **5** (11), 9154–9160.

80. Chen, L., Hernandez, Y., Feng, X., and Müllen, K. (2012) From nanographene and graphene nanoribbons to graphene sheets: chemical synthesis. *Angew. Chem. Int. Ed.*, **51** (31), 7640–7654.
81. Lu, J., Yeo, P.S., Gan, C.K., Wu, P., and Loh, K.P. (2011) Transforming C60 molecules into graphene quantum dots. *Nat. Nanotechnol.*, **6** (4), 247–252.
82. Cai, J., Ruffieux, P., Jaafar, R., Bieri, M., Braun, T., Blankenburg, S., Muoth, M., Seitsonen, A.P., Saleh, M., Feng, X., Müllen, K., and Fasel, R. (2010) Atomically precise bottom-up fabrication of graphene nanoribbons. *Nature*, **466** (7305), 470–472.
83. Son, Y.-W., Cohen, M.L., and Louie, S.G. (2006) Energy gaps in graphene nanoribbons. *Phys. Rev. Lett.*, **97** (21), 216803.
84. Han, M.Y., Özyilmaz, B., Zhang, Y., and Kim, P. (2007) Energy band-gap engineering of graphene nanoribbons. *Phys. Rev. Lett.*, **98** (20), 206805.
85. Müllen, K. (2014) Evolution of graphene molecules: structural and functional complexity as driving forces behind nanoscience. *ACS Nano*, **8** (7), 6531–6541.
86. Iyer, V.S., Wehmeier, M., Brand, J.D., Keegstra, M.A., and Müllen, K. (1997) From hexa-peri-hexabenzocoronene to "superacenes". *Angew. Chem. Int. Ed.*, **36** (15), 1604–1607.
87. Yang, X., Dou, X., Rouhanipour, A., Zhi, L., Rader, H.J., and Müllen, K. (2008) Two-dimensional graphene nanoribbons. *J. Am. Chem. Soc.*, **130** (13), 4216–4217.
88. Bieri, M., Treier, M., Cai, J., Aït-Mansour, K., Ruffieux, P., Gröning, O., Gröning, P., Kastler, M., Rieger, R., Feng, X., Müllen, K., and Fasel, R. (2009) Porous graphenes: two-dimensional polymer synthesis with atomic precision. *Chem. Commun.*, **45**, 6919–6921.
89. Bieri, M., Blankenburg, S., Kivala, M., Pignedoli, C.A., Ruffieux, P., Müllen, K., and Fasel, R. (2011) Surface-supported 2D heterotriangulene polymers. *Chem. Commun.*, **47** (37), 10239–10241.

3
Graphene Composites

Suman Chhetri, Tapas Kuila, and Naresh Chandra Murmu

3.1
Introduction

Nanomaterials form the core of modern science and technology and, as compact and light-weight materials, are in high demand in fields as diverse as aerospace to automotive, electronics to energy, environment to biomedical, and so on [1]. However, the design and development of efficient and high-performance materials is challenging as it requires understanding of the property, morphology, and dimension of existing materials as well as of the synthesis of new materials from it, with improved properties. Therefore, in order to cope up with the demand of modern day life, it is imperative to fabricate novel materials having possible applications. Thus, tailoring of a new material is very crucial to enhance the intrinsic properties of existing materials and to make it applicable to a wide spectrum of fields without compromising on environmental sensitivities.

Of late, combining various types of materials to achieve synergy in their properties has become the common practise in the realm of material science and technology. Polymeric materials have always been used as one component because of their easy availability in various scaffolds, low cost, easy fabrication, and broad areas of applicability. The integration of the second phase called filler or additives and the polymeric matrix results in the formation of a new material known as polymer composites [2]. If the filler dimension is in the range of a nanoscale, then the resultant material formed is called polymer nanocomposites [3]. The above described filler can be of zero-dimensional fullerenes, 1D carbon nanotube (CNT), 2D layered silicate or graphene, and 3D graphite [4–7]. The field polymer nanocomposites have been considered as promising materials in the paradigm of nanoscience and technology.

The earliest research on nanocomposites incorporating layered silicate as filler dates back to 1950 [8]; however, it could not create much hype in the scientific and industrial realm until Toyota Motor Corporation came up with their harbinger work in this area where they reported the enhancement in the mechanical property of Nylon-6 using montmorillonite as fillers [9]. This work stimulated plethora of research in the area of polymer composites by incorporating nano-dimension

Graphene Technology: From Laboratory to Fabrication,
First Edition. Edited by Soroush Nazarpour and Stephen R. Waite.

filler of different nature, geometry and size, to achieve high performance and multifunctional material. Conventional polymer composites usually employed discrete fillers such as glass fibre, metal particle, wood sheets, and so on [10]. Only modest improvement in various properties was observed in conventional polymer composites due to lower surface area of filler particles, mismatch in the properties of the individual components, incompatibility, and poor interfacial interaction/adhesion and difference in coefficients in thermal expansion [11]. For many years, nanoadditives such as clay or LDHs (synthetic clays), expanded graphite (EG), carbon nanofibre (CNF), CNT, carbon black, metal nanoparticles, and so on have remained candidates for vibrant fields of research due to their high surface area, good compatibility with matrix, and ability to tailor their surface to control the interaction [12–17]. The developed materials showed enhanced physicochemical properties and proved to be superior to conventional polymer composites [18–22].

The underlying principle behind the preparation of polymer nanocomposites using the above mentioned fillers is to couple the intrinsic potential of multiple components and to enhance the performance of the resulting materials. Further, small loadings of nanofillers are more effective in enhancing the performance of composites as compared to nanostructured filler, where its higher content makes the material brittle and heavy [23]. Owing to their layered structure and low cost, clay nanominerals (montmorillonite, MMT) are very effective in increasing the mechanical strength of the composites, but their low thermal electrical conductivity and stiff nature make it a less favorite for composite fabrication [24, 25]. Though, metal nanoparticles proved effective filler and it impart conductivity to the matrix otherwise insulator but its poor compatibility with the matrix make it less viable. Conductive carbonaceous nanomaterials such as EG, CNT, CNF, and carbon black are considered as favorable alternatives as filler material [26–29]. Among them, CNT nanofiller dominated the area as a promising candidate and is still seen as a potential filler because of its superb mechanical and thermal properties, low density, and high aspect ratio [10]. However, poor dispersion in and interaction with polymer matrix and high cost of production inhibited its candidature in the development of polymer composites [30]. The discovery of graphene in 2004 showed a new horizon in the path toward the development of advanced multifunctional composite materials [31, 32].

Graphene, one-atomic-thick sheet of sp [2] bonded carbon arranged in a two-dimensional lattice has drawn lot of interest from fields such as physics, chemistry, materials science, and medical science owing to its intriguing and exciting properties such as high surface area; large aspect ratio; outstanding mechanical, electrical, thermal properties; flexibility; high transparency; and optoelectronic properties to name a few [33–36]. Its high surface-to-volume ratio and light weight makes it superior to CNT in the manufacture of light weight material for aerospace and automotive applications. Thermal and electrical properties of graphene are higher than those of CNT albeit their mechanical properties are comparable. Further, graphite is the precursor of graphene and its derivatives; so, cost of production of graphene is low as compared to other carbon-based fillers.

Graphene as well as its derivatives has the potential to modify the properties and even has the ability to induce new properties to the matrix. Graphene/polymer composites show improved mechanical, thermal, barrier, and flame retardant properties [32, 37–40]. Hence, ever since its inception, graphene has been seen as a versatile candidate for the fabrication of polymer nanocomposites. However, difficulties lie in transforming or imparting the unique properties of graphene to the polymer. To get the utmost benefit of a filler in terms of properties, homogeneous dispersion of the filler in the matrix is most crucial with good interfacial interaction and adhesion between the phases [41]. Degree of enhancement in the properties of the resultant composites depends upon the extent of interfacial bonding between the host matrix and the filler. Individual graphene sheets show poor compatibility with the polymer, and therefore, it is crucial to exploit its extraordinary properties to make graphene compatible with and dispersible in the host matrix. Functionalization of graphene not only makes graphene compatible with the matrix but also increases its affinity toward the matrix and adhesion/interaction becomes strong [42]. Both covalent and non-covalent surface modification of GO have been investigated in detail using organic and inorganic surface-modifying agents.

This chapter tries to address a wide range of thermoplastic and thermoset polymers, biodegradable polymers, and so on Graphene nanoplatelets (GnPs), chemically functionalized graphene (CFG), GO, functionalized graphene oxide (FGO), thermally reduced graphene oxide (TrGO) and pristine graphene incorporated polymer composites have been highlighted in detail. The improved physiochemical properties of the graphene/polymer composites including their possible areas of applications have also been covered.

3.2
Preparation and Properties of Graphene

The emergence of graphene in 2004 mesmerized the scientific community because of its unique properties that could be exploited to generate new class of materials. Although graphene was first isolated by mechanical exfoliation of graphite [31], this method could not generate the amount of graphene required for various possible applications. The micromechanical exfoliation technique employed repeatedly peeling off pyrolytic graphite using scotch tape. The other synthetic methods that can supply graphene required for various applications are chemical vapor deposition (CVD) (also known as "bottom-up"), epitaxial growth, plasma enhanced CVD, unzipping of CNT, solution-based reduction of GO, and so on [31, 43–48]. CVD is one of the promising methods to synthesize mono- or few-layer graphene sheets using metal foil at over 1000 °C [49]. These preparative methods produce highly crystalline graphene sheets that are suitable for studying the fundamental properties of graphene but are not viable for composite preparation as it demands large amount of fillers. Moreover, CVD grown graphene is not compatible with the organic polymer matrix. Thus, for composite fabrication purpose, preparation of graphene from GO is most appropriate as its precursor graphite is abundantly available and can be engineered for scaled-up production. The route followed is oxidation of graphite to graphite oxide, which is exfoliated to GO layers and subsequently converted to RGO (reduced GO) sheets. Although, this is also not without flaws, reduction might not preserve all the intrinsic properties of pristine graphene. Modified Hummer's method is the most common method to prepare GO from graphite [50]. Most of the procedures to prepare graphene/polymer composites and graphene-like materials have used GO as the starting material. The oxidation process introduces various oxygen functionalities such as epoxide, hydroxyl, diol, ketone on the basal plane and carboxyl, and carbonyl on the edges of the surface and electronic conjugation of graphite gets disturbed/disconnected [51–53]. To generate RGO, typically exfoliation of graphite oxide to GO and its reduction by chemical, thermal, electrochemical, and so on are carried out [54]. The materials so formed are predicated to structurally resemble pristine graphene and said to have conserved its unique properties.

Graphene, an allotrope of carbon has the thickness of 0.35–1 nm [55]. The Young's modulus and Poisson's ratio for graphene are 1.02 TPa and 0.149, respectively [56]. Electrical and thermal conductivity of graphene is $6 \times 10^5\ S\ m^{-1}$ and $5.1 \times 10^3\ W\ m\ K^{-1}$, respectively [36,57]. The theoretical intrinsic carrier mobility ($200{,}000\ cm^2\ V^{-1}$) and specific surface area ($2600\ m^2\ g^{-1}$) of graphene are also very high as compared to the traditional reinforcing fillers [32, 58]. High aspect ratios and good mechanical flexibility make graphene an interesting material among the carbon allotrope members. Graphene being an aromatic moiety, it undergoes electrophonic reaction other than nucleophilic and cyclo-addition click reaction and so on [59, 60]. A reaction on the surfaces of graphene destroys the graphitic planar structure, disconnecting sp^2 network that eventually hampers electrical conductivity.

3.3
Functionalization of Graphene

The reinforcement ability of filler to matrix basically depends upon the extent of its dispersion and its effectiveness in transferring load at the interface. Property enhancement of composites only becomes pragmatic when the filler is uniformly distributed and is able to develop some affinity toward the matrix. The nature of the interface plays a crucial role in the performance of the composites. However, pristine graphene does not disperse in organic polymer and agglomerates due to Vander Waals force of attraction and π–π interaction thereby affecting the desired properties. Moreover, the hydrophobic nature of graphene is incompatible with the polymer matrix. Therefore, prior to incorporation into the matrix, the surface of graphene should be functionalized with surface modifying agents depending on the type of polymer matrix. Study shows two types of surface modification: (i) covalent and (ii) non-covalent [59]. In the first kind, small or large organic moieties usually grafted to the surface of GO through covalent bond and latter occurs through π–π interaction between the graphene surface and the benzene ring carrying organic species [60, 61]. In non-covalent functionalization, no bond formation takes place between surfactant and GO, so the interface is not much effective in load transformation as covalent functionalization. Although, covalent functionalization is a better technique to tune the interface, which eventually influences the load transfer efficiency it hampers the electrical properties.

3.3.1 Covalent Modification

Covalent modification can be carried out through nucleophilic substitution, electrophilic substitution, condensation, and cyclo-addition reaction. The functional groups such as porphyrins, poly(norepinephrine), oleylamine, and so on have been grafted onto the surface of GO [62–64]. Primary amines have been used to simultaneously functionalize and reduce GO [65]. In nucleophilic substitution reaction, the reactive epoxy groups of GO react with organic species bearing lone pair of electrons. This method is easy to process as it takes place at room temperature (RT) and produces scalable functionalized GO (fGO) required for the fabrication of graphene/polymer composites. Li *et al.* obtained simultaneously reduced and functionalized GO using octadecyl amine and incorporated into the polystyrene (PS) matrix. The fGO/PS composite showed enhanced mechanical and electrical properties at very low fGO loadings [66]. Kuila *et al.* prepared octadecyl amine functionalized graphene and fabricated ethylene vinyl acetate (EVA) co-polymer composites with enhanced mechanical properties and thermal stability [67]. Jang *et al.* modified the surface of GO using three types of alkyl amine namely, octylamine (OA), dodecylamine (DA), and hexadecylamine (HDA) to increase dispersibility in low-polar solvents and incorporated into PS [68]. It is seen that the properties of the composites were significantly affected by alkyl chain length. The thermal stability and mechanical properties of the composites with longer alkyl chain amine–modified GO was higher as compared to the composites with short chain alkyl amine functionalized GO. Kim *et al.* reported ternary roles of ethylenediamine to functionalize, reduce, and stitching of GO and, so, formed conductive RGO was incorporated into the linear low density polyethylene (LLDPE) matrix [65]. The composites showed enhanced mechanical properties suggesting the effective reinforcement of LLDPE with stitched RGO.

One of the widely practiced and effective approaches other than nucleophilic substitution to achieve FG is amidation reaction, where edge –COOH groups of GO attached with the $-NH_2$ or cyanate groups of the modifiers form amide linkage. Stankovich *et al.* used organic isocyanates to functionalize GO which formed a stable dispersion in polar aprotic solvents [69]. Figure 3.1 shows the proposed reaction mechanism between the organic isocyanates and GO showing the formation of amide and carbamate ester linkage. Many $-NH_2$ terminated molecules have been attached to GO surfaces intended for various end

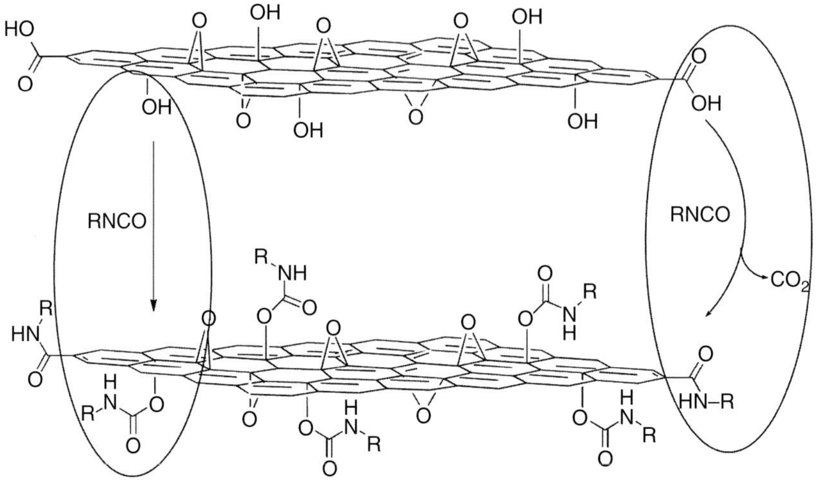

Figure 3.1 Proposed reactions during the isocyanate treatment of GO, where organic isocyanates react with the hydroxyl (left oval) and carboxyl groups (right oval) of graphene oxide sheets to form carbamate and amide functionalities, respectively. (Reproduced from Ref. [69] with permission from Elsevier.)

applications. Amine terminated polyethylene glycol (PEG) reacted with GO platelets through an amide bond and showed potential applications in drug delivery [70]. Biocompatible and water dispersible polysaccharides such as chitosan has been grafted onto the surface of GO by amidation reaction, which is useful in gene and drug delivery system [71].

Esterification is a widely used method to tune the surface GO. The edge –COOH groups of GO react with CH_2OH terminated molecules. Salavagione *et al.* covalently functionalized graphene with poly(vinyl alcohol) (PVA) and polyvinyl chloride (PVC) through esterification reaction [72, 73]. The esterification reaction of PVC with GO was carried out in the presence of DCC/DMAP followed by $NaBH_4$ reduction. PVA grafting was carried out by two methods: first by using DCC/DMAP and second by using $SOCl_2$ activated GO followed by reduction. Zhang *et al.* used fullerene to functionalize the surface of graphene [74]. The attachment of fullerene to graphene surface restrains graphene sheet piling and helped in exfoliation.

The GO can be covalently functionalized with silane through hydrolytic condensation. Wang *et al.* functionalized graphene nanosheets (f-GNSs) by grafting 3-aminopropyl triethoxysilane [75]. The silane-functionalized graphene was then incorporated into the epoxy matrix, which showed significant improvement in thermal and mechanical properties. Addition of 1 wt% of functionalized graphene nanosheets (f-GNSs) increased the thermal degradation temperature by ~20 °C and the tensile strength by 45% as compared to neat epoxy. Figure 3.2

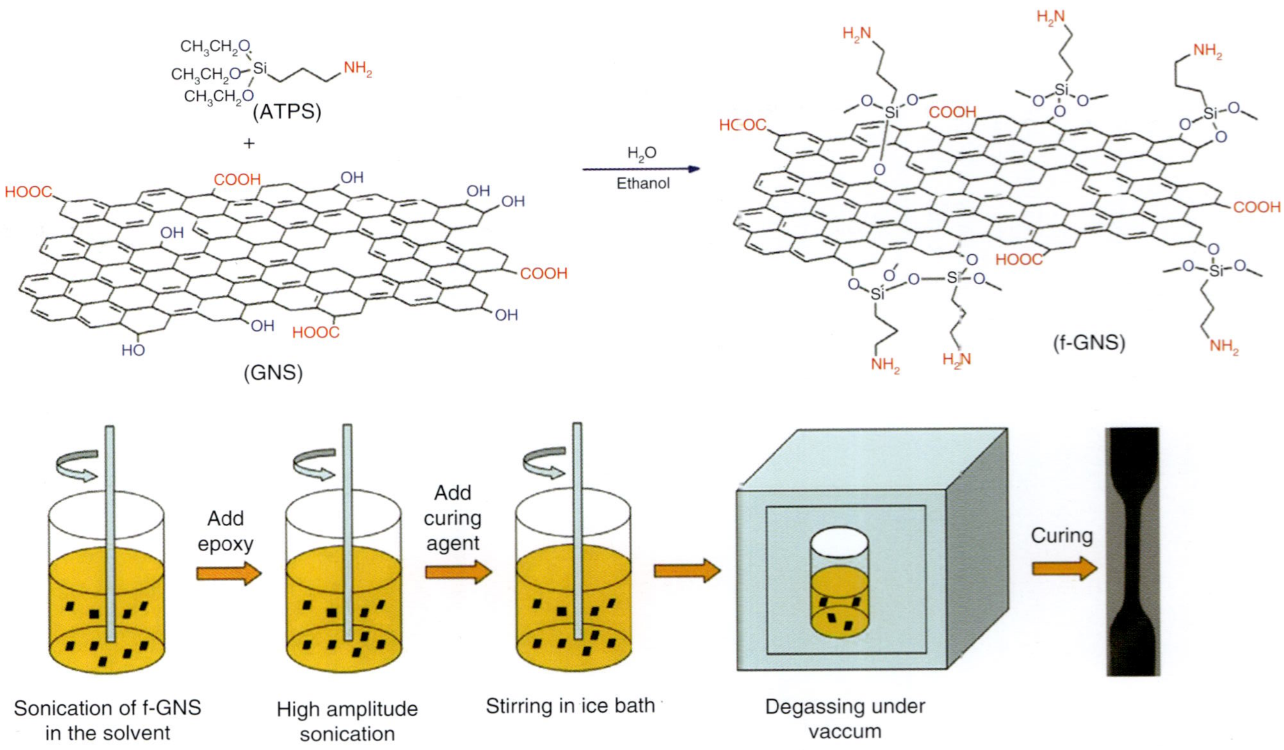

Figure 3.2 Schematic representation of functionalization and composite fabrication. (Reproduced from Ref. [75] with permission from Elsevier.)

shows the schematic for the functionalization and fabrication of f-GNSs/epoxy composites.

The sp^2 carbons of graphene undergo free radical addition reaction and react with dienophiles. The reactive free radicals of diazonium salt would attack the sp^2 bond of graphene and form a covalent bond [61]. During this reaction, sp^2 hybridized carbon transforms into sp^3, and the aromatic system gets disturbed, which resulted in a remarkable decrease in electrical conductivity. Fang *et al.* showed that the hydroxyl aryl groups can be grafted on the surface of graphene via diazo-coupling reaction and at the same time they can act as initiators for the polymerization of styrene via the atomic transfer radical polymerization (ATRP) method [76]. It has been found that the diazonium salt of benzoyl peroxide can also be used for the surface modification of graphene via free radical addition reaction. Kan *et al.* fabricated graphene grafted PS–polyacrylamide (PS–PAM) copolymer of styrene and acyl amide using benzoyl peroxide as initiator [77]. Addition of dienophiles to graphene is also a common method in the addition reaction category of graphene besides free radicals. Azomethine ylide is the most common dienophiles used in the functionalization of carbonaceous nanostructure [61]. It reacts through 1,3 dipolar cyclo addition reaction and can offer a range of organic derivatives, which shows utility in a wide area of applications.

The surface of graphene can be functionalized by polymers using surface-initiated ATRP and this technique generates surface controlled graphene that is required for various applications. In the ATRP technique first monomers and the initiators are adsorbed to the hydroxyl or carboxylic acid groups of GO followed by the polymerization of the monomers to polymer chains. Lee *et al.* carried out polymerization of styrene, methyl methacrylate (MMA), and *t*-butyl acrylate (tBA) via ATRP on the GO surface [78]. The polymer-functionalized graphene formed stable dispersion in DMF, chloroform, toluene, and dichloromethane. Layek *et al.* grafted PMMA on the surface of graphene by ATRP technique to prepare FG [79]. The PMMA-functionalized graphene (MG) showed good dispersion in a poly(vinylidene fluoride) (PVDF) matrix, which resulted in an increase of 124% storage modulus, stress at break 157%, and Young's modulus 321% for 5% MG and 21 °C increase in T_g.

3.3.2 Non-Covalent Modification

The non-covalent surface modification with surfactants, macromolecules, ionic liquids, and aromatic moieties are of paramount interest because these techniques tune the surface of graphene keeping its intrinsic properties intact. The interaction that involved is $\pi-\pi$, hydrogen bonding and hydrophobic attraction [60]. Non-covalent technique has been widely exploited to achieve the stable dispersion of graphene in various solvents. Largely, $\pi-\pi$ stacking interaction and hydrogen bonding are most common as it concurrently stabilizes the graphene colloidal suspension and maintains the integrity of graphene. However, no bond formation takes place in non-covalent mode of functionalization; transfer of stress may not be as effective as in covalent functionalization. Stable aqueous dispersion of graphene has been prepared using water-soluble preen derivative, 1-pyrenebutyrate (PB^-), as a stabilizer because preen can interact with the graphene sheet through π−π interaction [80]. Large area flexible graphene film with electrical conductivity of $200\,S\,m^{-1}$ was obtained by filtration of this dispersion [80].

Graphene can also be non-covalently functionalized with polymers through hydrogen bonding or π−π interaction. Liang *et al.* used simple water solution processing technique to prepare GO/PVA composites and they reported a ~76% improvement in tensile strength and a 62% improvement in Young's modulus with the addition of only 0.7 wt% of GO [81]. Sulfanilic acid azocromotrop (SAC) was used to non-covalently functionalize GO through π−π interaction between the graphene and aromatic moiety of SAC [82]. Teng *et al.* functionalized GNSs non-covalently through π−π stacking of pyrene molecules with poly(glycidyl methacrylate) functional segment polymer chains (Py-PGMA) [83]. Py-PGMA contains oxirane rings that may enhance the compatibility between GNS and epoxy matrix resulting in enhanced thermal conductivity of the composites. Sulfonated poly-ether-ether-ketone interacts with graphene sheets through π−π interaction and formed FG sheets (SPG) [84]. The SPG dispersed homogeneously in the PVDF through the dipolar interaction and formed high-performance composites [85].The Young's modulus of the composites increased ~160% and the oxygen permeability coefficient decreased by ~91% with the addition of only 3 wt% of SPG, as compared to neat PVDF.

3.4 Preparation of Graphene Polymer Composites

The processing condition of composites influences the end properties of graphene-based polymer composites. The interfacial specific surface area, dispersion, interaction/adhesion, distribution of filler, and so on determine the overall properties of the composites. Graphene/polymer composites can be prepared by following two different routes, namely, mixing and template methods (in situ polymerization). Mixing methods include solution while melt methods include compounding technique.

3.4.1 In Situ Polymerization

Interaction between the filler and matrix facilitates the load transfer across the interface and plays a pivotal role in maximizing composites properties. Preparative methods influence the dispersion and hence interaction; in situ polymerization is one of the effective techniques to homogenously disperse graphene and its derivatives in polymer matrix [86]. In this technique, at first graphene or its derivatives is dispersed into the monomer followed by polymerization with the help of heat or radiation [86, 87]. Subsequently, polymer is supposed to be grafted onto the surface of graphene or its derivatives resulting in fine dispersion and good adhesion between the two components. Wide range of polymer composites such as graphene/PMMS, graphene/PS, LDH/polystyrenesulfonate (PSS), LDH/polyimide (PI), and so on have been prepared by this method [88–93]. Potts *et al.* prepared graphene/PMMA composites by in situ polymerization and studied the thermo-mechanical properties of the composites [94]. The composite showed ~15 °C rise in T_g and storage modulus was also improved. Lee *et al.* prepared functionalized graphene sheets (FGS)/waterborne polyurethane (WPU) composites by in situ methods [95]. The finely dispersed FGS particle formed a conducting network and, thus, conductivity of the composite improved. Further, improvement in storage modulus and crystallization of the soft segment of WPU were reported.

3.4.2 Solution Mixing

In this method, graphene is first dispersed in a solvent (aqueous and organic) and then mixed with polymer assisting by sonication, mechanical stirring followed by evaporating the solvent [96]. Although, this technique is found to be effective for fine dispersion of filler, use of toxic and hazardous solvent and complete removal of solvent are major issues. Wide range of polymer composites such as graphene/PVA, EVA/LDH, graphene/PMMA, graphene/PU, and so on have been processed by solution mixing technique [81, 97–99]. Goncalves *et al.* prepared PMMA grafted GO (G-PMMA) via ATRP and then the composites of G-PMMA/PMMA were fabricated [98]. The grafting of PMMA chain onto the GO helped to make good interaction with the bulk PMMA resulting in increase in elongation at break and in the films becoming stiffer as compared to the neat PMMA. In addition, G-PMMA effectively reinforced thermal stability of the composites. Kuila *et al.* prepared DA-G/LLDPE composite by solution mixing and studied the mechanical, thermal, and electrical properties [100]. The composite showed ~46% increase in tensile strength with 3 wt% DA-G loadings. The onset degradation temperature of the composite (3 wt% of DA-G) was increased by ~40 °C as evidenced from thermogravimetric analysis (TGA).

3.4.3 Melt Mixing

Melt mixing technique is highly efficient when polymers do not go into solution or are inappropriate for in situ polymerization. Melt blending technique has been found to be suitable especially to fabricate thermoplastics composite materials. The principle behind the process is that first polymers are heated to form a viscous melt and then high shearing forces are applied to disperse graphene particles using extrusion and injection molding. This technique is eco-friendly and mass production of composite material intended for various applications can be possible as it does not involve solvent. Further, the properties of composites can be controlled by varying extruder speed, processing temperature, and duration of mixing. Despite these advantages, melt blending process suffers weaknesses, as high temperature applied during compounding may degrade the polymers and even the filler materials. The high shear force applied for dispersion may also break graphene sheets or may roll its surface, which eventually hampers the functional properties of the composites. Polypropylene (PP), polylactide (PLA), polyethyleneterephthalate (PET), and so on have been widely used to fabricate composite following melt mixing [101–103]. Song *et al.* fabricated PP latex coated graphene/PP composites by melt mixing and reported ~75% increase in yield strength, ~74% increase in Young's modulus, and 2.5 °C rise in T_g at only 0.041 vol% of PP latex coated graphene loadings [104]. Lin *et al.* prepared polyethylene (PE) grafted GO/PE composites by melt mixing method [105]. They coated 3-aminopropyltriethoxysilane (APTES) onto the GO sheets and subsequently maleic anhydride was grafted onto it. Coating of APTES was confirmed by X-ray photoelectron spectroscopy (XPS) and Fourier transform infrared (FT-IR) spectroscopy. They observed that the crystallization temperature of the maleic anhydride PE grafted GO was ~10 °C higher than that of pure maleic anhydride grafted PE. The improvement in Young's modulus (0.3% modified GO loading) and tensile strength (0.1% modified GO) was recorded as 20% and 15%, respectively.

3.4.4 Other Preparative Technique

Layer-by-layer (LbL) assembly is one of the fruitful techniques where graphene component and polymer can be alternately assembled in layers [106]. This technique offers fine dispersion and orientation of graphene in the final composites. It can also control the distribution, amount of graphene, and morphology of the composites [10, 106]. The technique that is used to deposit the layers are spin- or spray-assisted LbL method, vacuum-assisted method, and dip-assisted LbL method. Very few reports are available, which have followed the LbL technique to fabricate film composites. Zhao *et al.* reported the preparation of exfoliated GO/PVA multilayer film through hydrogen bonding and studied its mechanical properties [107]. Kulkarni *et al.* reported multilayer ultrathin GO/polyelectrolyte by spin-assisted LbL assembly accompanied by Langmuir–Blodgett deposition and reported enhanced mechanical properties and elastic modulus [108].

Supramolecular arrangement of layers through interaction such as hydrogen bonding is also one of the techniques to tune the interface and orient the layers precisely. Layek *et al.* anchored sulfonate groups onto the surface of graphene to improve its dispersion in the PVA matrix [109]. The fibrillar, dendritic, and rod-like structures formed due to hydrogen bond formation showed improved tensile strength, strain at break, toughness, and Young's modulus. Further, they revealed ~10 orders of increase in DC conductivity of the composites over neat PVA. Usually, graphene/PVA composites have shown improvement in tensile strength at the expanse of strain at break. All these observations throw interesting insight that the preparative method plays a crucial role in the improvement of the properties of composites.

3.5
Characterization of Graphene-Polymer Composites

Graphene is able to deliver its unique properties only in its single layer form; its restacking and aggregates pose serious problem in the processing and fabricating of polymer composites. So, it is important to characterize graphene-incorporated polymer composites to know the degree of exfoliation and dispersion. The structural characterization of graphene/polymer composite can be achieved by X-ray diffraction (XRD), UV–Vis spectroscopy, FT-IR spectroscopy, XPS, and Raman spectroscopy [110]. The extent of dispersion and morphology of the filler can be examined by scanning electron microscopy (SEM) and transmission electron microscopy (TEM). The complete exfoliation of graphene sheets inside the polymer matrix can be traced by XRD as 3D graphitic structure destroys in this in few-layer graphene. The characteristic peak of graphite at $2\theta \sim 26°$ (002) shifts toward a lower angle ($2\theta \sim 10-11°$) in GO due to the increase in interlayer spacing. The interlayer spacing decreases and appears as a low intense broad hump at $2\theta = 25-26°$ in the RGO. So, XRD is a very helpful tool to know the extent of exfoliation of graphene in the matrix. The diffraction pattern of exfoliated graphene/PP latex composites show only the peak characteristics of PP indicating complete exfoliation of graphene in the matrix. Figure 3.3 shows the XRD pattern of graphite, graphite oxide, graphene, and exfoliated graphene PP composites [104].

FT-IR spectra are very informative to understand the oxidation, reduction, and functionalization of graphene sheets. The pristine graphite shows characteristic peak at 3430 and 1610 cm^{-1} and correspond to the characteristic of O–H

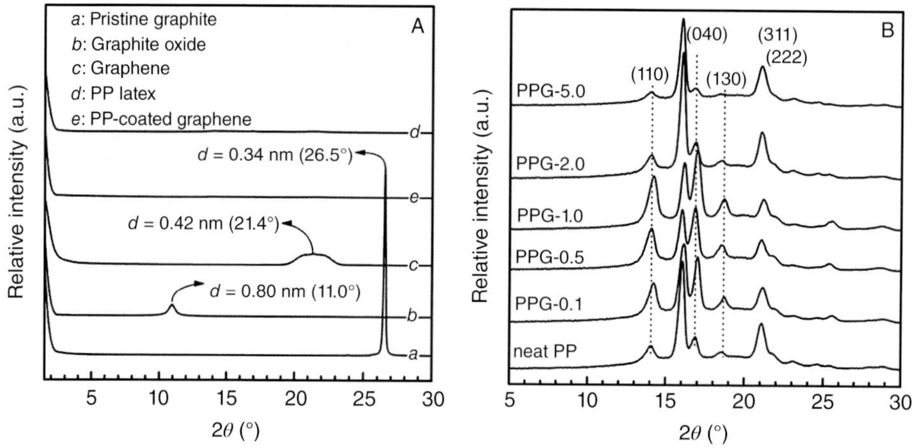

Figure 3.3 (A) XRD patterns of (a) pristine graphite, (b) graphite oxide, (c) graphene, (d) PP latex, and (e) PP-coated graphene. (B) XRD pattern of PP/graphene nanocomposites with various graphene loadings. (Reproduced from Ref. [104] with permission from Elsevier.)

stretching and C=C vibration, respectively. The peaks at 1728 and 1052 cm^{-1} are attributed to the C=O and C–O–C stretching, respectively. RGO shows peak at 3430 cm^{-1} but, with reduced intensity, the characteristic peak of epoxide and carboxyl disappear. DA-G/LLDPE composites show peaks at 2950, 2928, and 2857 cm^{-1} corresponding to the C–H stretching vibrations of -CH_3, -CH_2 and -CH groups, respectively. The peaks at 1470 and 1375 cm^{-1} (deformation of CH_2 and CH_3 groups) of neat LLDPE shifted to 1466 and 1370 cm^{-1} in the composites suggesting the interaction between LLDP and DA [111]. Hence, FT-IR is helpful to confirm the modification of graphene and may even help to check the interaction between the filler and matrix. Figure 3.4 shows the FT-IR spectra of GO, DA-G, and its composites with LLDPE [111].

The SEM image of the fracture surface of composites gives insight into the dispersion and distribution of filler in the polymer matrix. It is very helpful to check the interaction between the filler and matrix. Figure 3.5 shows the FESEM micrographs of sulfonated graphene (SG)/ poly(vinyl alcohol) (PVA) composites with dendritic and rod like structures at different compositions. The appearance of different structures with composition may be due to the difference of interfacial interaction between SG and PVA. TEM is a very helpful technique to examine the distribution of graphene sheet in the polymer matrix. It is also helpful to determine the number of graphene layers; as the thickness of the layers is critical to properties improvement, it is imperative to know of what thickness sheets are distributed in the matrix. Figure 3.6 shows the uniform distribution of dodecyl-modified graphene in LLDPE matrix prepared by Kuila *et al.* [100].

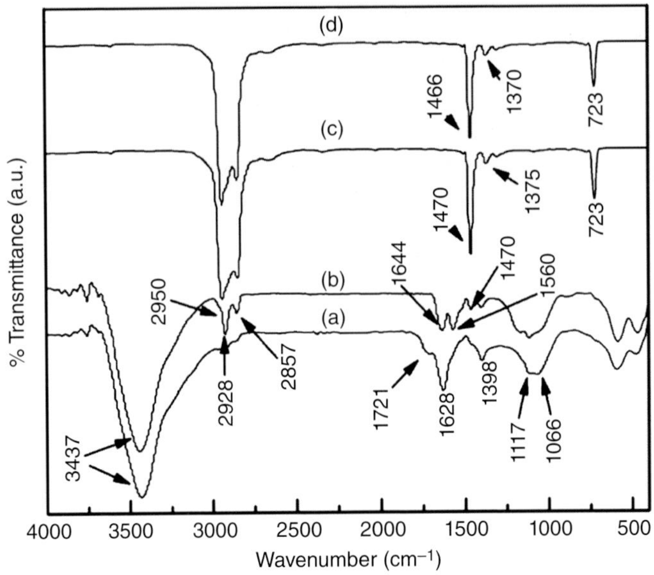

Figure 3.4 FTIR spectra of (a) graphite oxide, (b) DA-G, (c) LLDPE, and (d) LDG-3. (Reproduced from Ref. [111] with permission from Elsevier.)

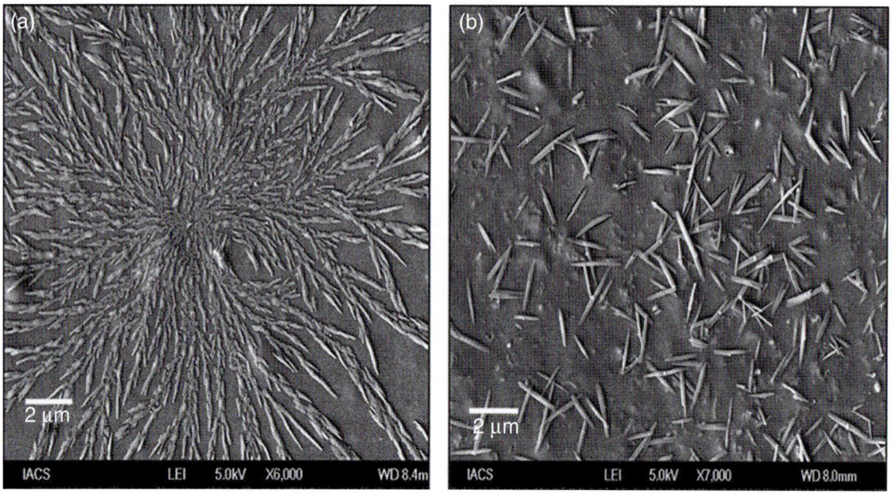

Figure 3.5 FESEM micrographs of (a) dritic and (b) rod like structured PVA/SG composites. (Reproduced from Ref. [109] with permission from Elsevier).

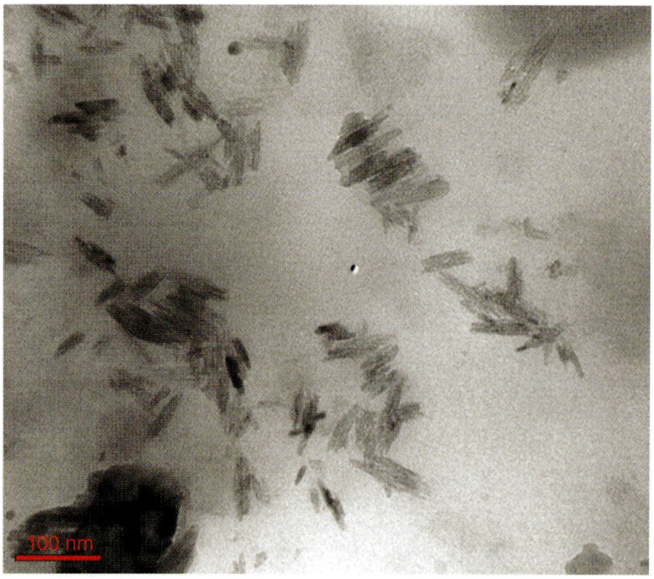

Figure 3.6 TEM image of DA-G/LLDPE composite. (Reproduced from Ref. [100] with permission of Elsevier.)

3.6 Properties of Graphene/Polymer Composites

Graphene is the strongest known material with many astonishing properties that can be useful in many applications. It has the ability to significantly improve various properties of the composites at very low loadings. Graphene and its derivatives have been extensively studied in relation to polymer composites, energy storage, and biology, and flame- retardant properties to name a few. With the development of advanced processing techniques, graphene is getting more attention as a promising material for future generation technology.

3.6.1 Mechanical Properties

Graphene has exceptional mechanical properties, namely, high Young's modulus, high aspect ratio, large tensile strength, and flexibility, which make it suitable for reinforcing the physicochemical properties of neat polymer. Although pristine graphene is difficult to disperse in polymeric matrices due to its hydrophobic nature, suitably functionalized graphene is very effective in reinforcing the mechanical properties. The degree of enhancement in the mechanical properties of the composites depends upon two critical factors: (i) the dispersion of filler in the composite, which determines the specific interfacial area and (ii) the interfacial strength, which administrates the stress transfer across the interface. Both the factors are very crucial but interfacial strength plays a major role that ultimately decides the mechanical properties. Interface can be designed by modifying the surface of graphene, which also helps to disperse it in the organic solvent and prevents restacking of layers. Thus, homogenous dispersion, proper exfoliation of graphene materials, and a strong interface are crucial to improve the mechanical properties.

However, fabrication of composites with enhanced mechanical properties is cumbersome. Fine dispersion of nanofillers into the matrix, large interfacial area, and strong interface decides the overall mechanical properties of the composites. If any kind of interaction such as covalent, Vander Waals interaction, $\pi-\pi$ interaction, hydrogen bonding, and so on exists across the interface, then composites with enhanced mechanical properties can be anticipated. Such types of interactions play a pivotal role in strengthening the interface, as strong interface facilitates the stress transfers from nanofiller to the host matrix, thereby, making composites more robust. Graphene and its derivatives have been considered as a successful filler in improving composite's mechanical properties; however, there are a few studies that report improvements in mechanical parameters such as tensile strength, elastic modulus, elongation, and toughness. It is noted that the addition of filler may be helpful for the improvement of a certain property but may cause the deterioration of other properties. Therefore, careful optimization of fillers (GNPs, rGO), dispersion, and interfacial strength is paramount to fabricate multifunctional high-performance composites. Wang *et al.* fabricated graphene-reinforced

PVA composites film by deploying GO and graphene into an aqueous solution of PVA [113].The enhancement in tensile strength and elongation at break was 212% and 34%, respectively with ~0.5 wt% of graphene content. Figure 3.7 shows the stress–strain curve of GO/PVA and graphene/PVA composites.

Efficient load transfer takes place across the interface of materials only when the dispersion of fillers in polymer and the interaction between the two components are optimum. Liang *et al.* accomplished a simple and eco-friendly technique to fabricate graphene/PVA composites [81]. The hydrogen bond formed between graphene and PVA helped in stress transfer across the interface. About 76% increase in tensile strength and a 62% improvement of Young's modulus by the addition of only 0.7 wt% of GO was recorded. Cano *et al.* covalently functionalized GO with six different molecular weights of PVA through esterification of carboxylic groups of GO with hydroxyl groups of PVA [114]. The PVA functionalized GO (f-PVA-GO) was prepared through vacuum filtration and showed enhanced mechanical properties than the composites filled with GO. The f-PVA-GO/PVA composite showed ~60% improvement in Young's moduli and ~200% improvement in tensile strength relative to GO filled PVA composites. Figure 3.8 shows the SEM images of the fracture surfaces of vacuum filtered paper-like materials of pure GO and f-PVA-GO.

Vadukumpully *et al.* prepared surfactant-wrapped graphene nanoflakes and PVA ultrathin composite films by a simple solution blending, drop casting, and annealing route [115]. The reinforcing ability of graphene improved the properties of the composites by ~58% in Young's modulus and 130% in tensile strength. Zhao *et al.* prepared fully exfoliated GNS/PVA composites through a facile solution mixing [116]. The composite with 1.8 vol% of graphene loadings showed ~150% improvement in tensile strength and ~10 times increase in Young's modulus as compared to neat PVA. In another study, poly(dopamine)-modified GO/PVA ("dGO/PVA") was prepared through solution mixing [117]. The composites showed improvement in tensile modulus, ultimate tensile

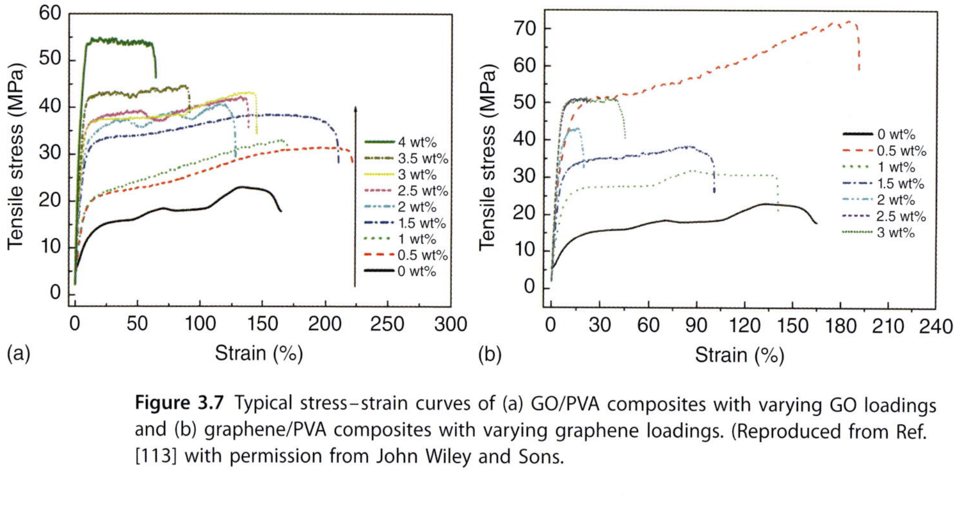

Figure 3.7 Typical stress–strain curves of (a) GO/PVA composites with varying GO loadings and (b) graphene/PVA composites with varying graphene loadings. (Reproduced from Ref. [113] with permission from John Wiley and Sons.

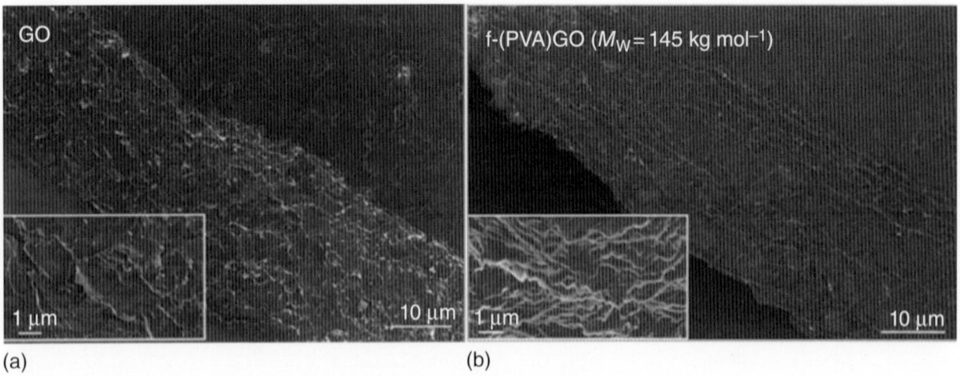

Figure 3.8 SEM images of (a) GO and (b) f-PVA-GO composites. The inset shows high resolution image. (Reproduced from Ref. [114] with permission from Elsevier.)

strength, and strain-to-failure by 39%, 100%, and 89%, respectively with 0.5 wt% dGO loadings [117]. It may be due to the combined hydrogen bonding interaction among GO, poly(dopamine), and PVA. Istrate *et al.* prepared graphene/PET by melt compounding and reported enhanced mechanical properties at 0.07 wt% of graphene loadings [118]. About 23% and 42% improvement in Young's modulus and tensile strength were recorded for the composites with 0.08 wt% filler loadings. The improvement in mechanical properties of the composites at very low loadings is industrially viable and may pave the way for graphene-based composites from laboratory to commercial production. In another study, Shim *et al.* prepared functionalized GO (f-GO) with alkylbromide of different alkyl chain length (hexyl, octyl, and dodecyl) by SN_2 reaction and used as filler to prepare PET composites [119]. The f-GO/PET composite showed improvement in tensile strength, Young's modulus, and elongation at break with the addition of f-GO. Figure 3.9 shows the tensile, Young's modulus and elongation-at-break curve of GO/PET and fGO/PET composites against filler loadings.

Liao *et al.* functionalized GO by octa(aminophenyl)silsesquioxane (OAPS) and incorporated into a PI matrix to study the reinforcement ability of GO/OAPS in the matrix by in situ polymerization [120]. The authors revealed that the results showed ultra-strong mechanical properties and low dielectric constant of the resultant composites. It may be ascribed to the strong interfacial interaction generated between OAPS and PI through covalent bonding. About 11.2-fold increase in tensile strength and a 10.4-fold improvement in tensile modulus against neat PI were observed, which are very encouraging to explore and fabricate mechanically robust light weight composite with low dielectric properties for various applications. Figure 3.10a,b shows the TEM image of GO and OAPS-GO at different magnifications and stress–strain curve of OAPS-GO film at different loadings of OAPS. Wang *et al.* prepared octadecylamine (ODA) functionalized GO (ODA-GA)/PI composites [121]. The composite with 0.3 wt% of ODA-GA loadings exhibited ~6.4-times increase in tensile modulus and 240% enhancement in tensile strength. Further, the composite showed decreases in

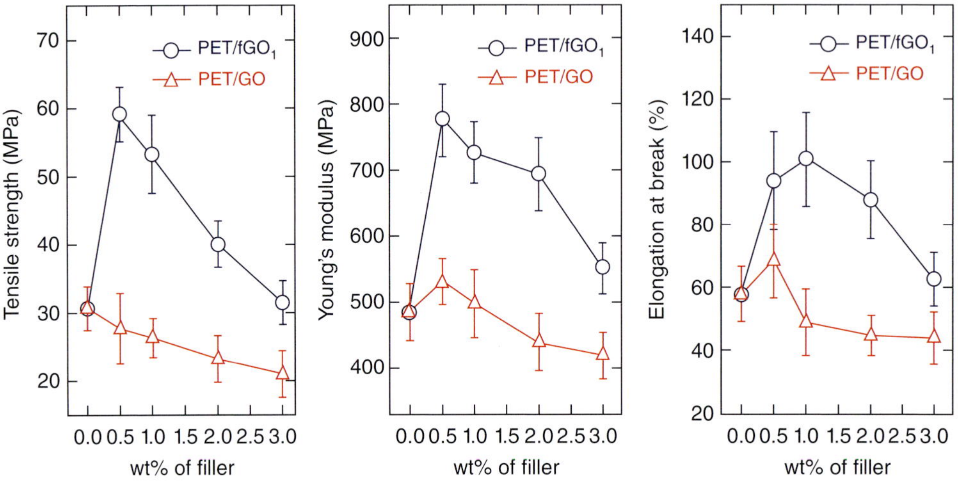

Figure 3.9 Typical stress–strain, Young's modulus and elongation-at-break cure of PET composites with GO and f-GO. (Reproduced from Ref. [119] with permission from the American Chemical Society.)

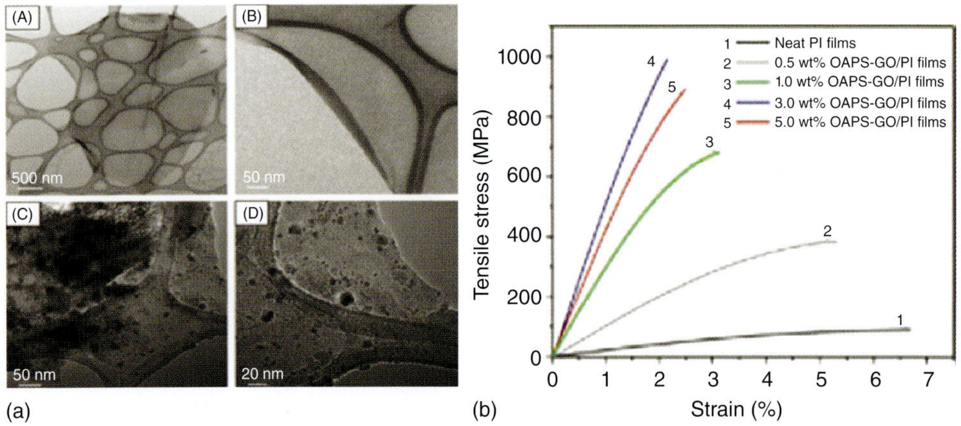

Figure 3.10 (a) TEM images of GO, OAPS-GO at low and magnification. (b) Typical stress–strain curve of OAPS-GO/PI film at various loadings of OAPS. (Reproduced from Ref. [120] with permission from the American Chemical Society.)

dielectric constant with the increase in ODA-GA content. The unprecedented rise in mechanical properties may be due to the fine dispersion of polymer grafted GO and its bonding interaction with PI. Liao *et al.* prepared graphene/PI composites that exhibited ~7.4 times increase in Young's modulus and 240% improvement in tensile strength with only 0.3 wt% graphene loadings [122].

Cai *et al.* prepared GO/PU composites and investigated the mechanical, thermal, and scratch resistance properties [123]. The improvement in mechanical properties can be ascribed to the good interface formed due to the covalent interaction between GO and PU. The composite showed ~7 times improvement in modulus as compared to neat matrix and ~50% toughness was achieved with 1% of GO loading. These findings indicate that GO is capable of reinforcing the properties of the composites and opens up many avenues to ponder upon the fabrication of engineering composites for probable applications. Xiang *et al.* fabricated hexadecyl-functionalized low-defect graphene nanoribbons (HD-GNRs)/TPU composite film by solution casting and investigated their mechanical and barrier properties [124]. The finely distributed HD-GNRs within the PU matrix resulted in phase separation of the TPU as evidenced from the FT-IR spectra. The integration of HD-GNRs with the TPU matrix enhanced mechanical properties and is presented in Figure 3.11. Wang *et al.* Prepared GNS/PU composites and it showed ~239% increase in tensile as compared to neat PU [125].

Rafiq *et al.* reported the preparation of FG filled nylon-11 and nylon-12 composites by direct melt compounding method with the assistance of premixing [126]. The authors studied the effects of FG on the ultimate tensile strength, elongation at break, impact strength, toughness, and permeation resistance of FG/nylon-11 and FG/nylon-12 composite. It is seen that the reinforcing ability of FG was very good at lower loadings. The FG/nylon 12 composites with 0.6 wt% of FG loadings exhibited improvement in ultimate tensile strength, elongation at break, fracture toughness, and impact failure energy by 35%, 200%, 75%, and 85%, respectively as compared to neat nylon 12. On the contrary, the FG/nylon 11 composites displayed impressive improvement of 250% in impact strength at the loading of 1 wt% FG but only slight improvement was recorded in tensile properties and fracture toughness. Roy *et al.* prepared polyethylenimine (PEI) functionalized GO (GO-PEI)/nylon composites by melt mixing technique. The improvement in tensile strength was 37% and 54% for 0.25 and 0.35 wt% of GO-PEI loadings, respectively.

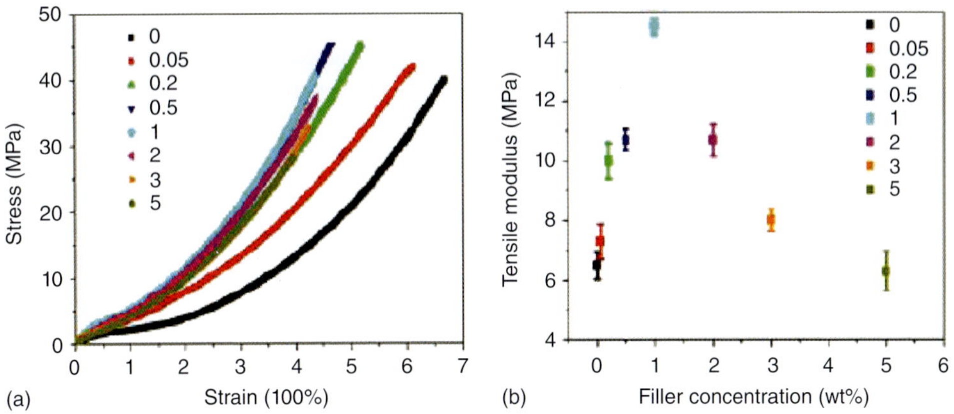

Figure 3.11 (a) Stress–strain curves of TPU and TPU/HD-GNRs composite films. (b) Summary of tensile moduli of different samples. (Reproduced from Ref. [124] with permission from the American Chemical Society.)

The improvement in Young's modulus of the composite was found to be 65% and 74% as compared to that in neat nylon [127]. The multiple reactive amine functional groups of PEI helped in dispersion and in the improvement in mechanical properties of the nylon matrix.

The inner surface of CNT is not accessible to make contact with polymer matrix, thus, affecting the reinforcing ability of CNT. Rafiee *et al.* unzipped multiwalled CNT to graphene nanoribbons and incorporated into the epoxy matrix to study the mechanical properties of the composites [112]. The improvement in Young's modulus and ultimate tensile strength at 0.3 wt% loading was found to be 30% and 22%, respectively, as compared to neat epoxy matrix. It may be attributed to the increased adhesion between the components after unzipping of cylindrical interfacial zone into a layered 2-D structure facilitating better load transfer at the interface. The effect of reinforcing ability of graphene platelets and single- and multiwalled CNT on the mechanical properties of epoxy nanocomposites has been investigated systematically [128]. Young's modulus, tensile strength, and mode I fracture toughness of the graphene/epoxy composites were 31%, 40%, and 53% higher than those of CNT/epoxy composites. It may be attributed to the higher platelet orientation of graphene as compared to that of CNT. Wajid *et al.* employed two different processing techniques, namely, solution mixing and freeze drying for the incorporation of polyvinylpyrrolidone (PVP) stabilized pristine graphene into the epoxy matrix [129]. Although, both the techniques are promising in terms of dispersion of stabilized graphene in the epoxy matrix, freeze drying/mixing was relatively more versatile in terms of graphene dispersion. It is seen that PVP not only stabilizes the graphene but also helps to create a strong interface with epoxy matrix through its polar lactam rings and load transfer across the matrix. The composite exhibited ~38% improvement in tensile strength and ~37% in Young's modulus at the loading of 0.46 wt%.

Recently, efforts have been made to assemble graphene sheets in 3D scaffold and find out its applications in different fields such as the automobile, aircraft, and marine industries, as well as in the manufacture of fire retardant materials. Its superb performance as filler material in the fabrication of composites has started gaining attention due to its unique 3D structure and excellent properties. The techniques that are being used to prepare 3D graphene nanostructure are the template-directing method, cross-linking method, in situ reduction assembly method, and so on [130–132]. Ni *et al.* prepared 3D graphene skeleton (3DGS) using poly(amidoamine) (PAMAM) dendrimer as reductant and cross-linking agent by self-assembly and in situ reduction method, and incorporated into the epoxy matrix [133]. In order to obtain fine dispersion and to arrange the graphene sheet in the matrix they adopted resin transfer molding method rather than direct mixing. The unique arrangement of graphene sheets and better dispersion resulted in ~120.9% and 148.3% increase in tensile and compressive strength, respectively. These results signify that if graphene sheets are arranged architecturally in a 3D structure, then it will endow better reinforcement and noble high-performance material can be anticipated. Figure 3.12 shows the variation of tensile, flexural, and compressive strength against filler loading.

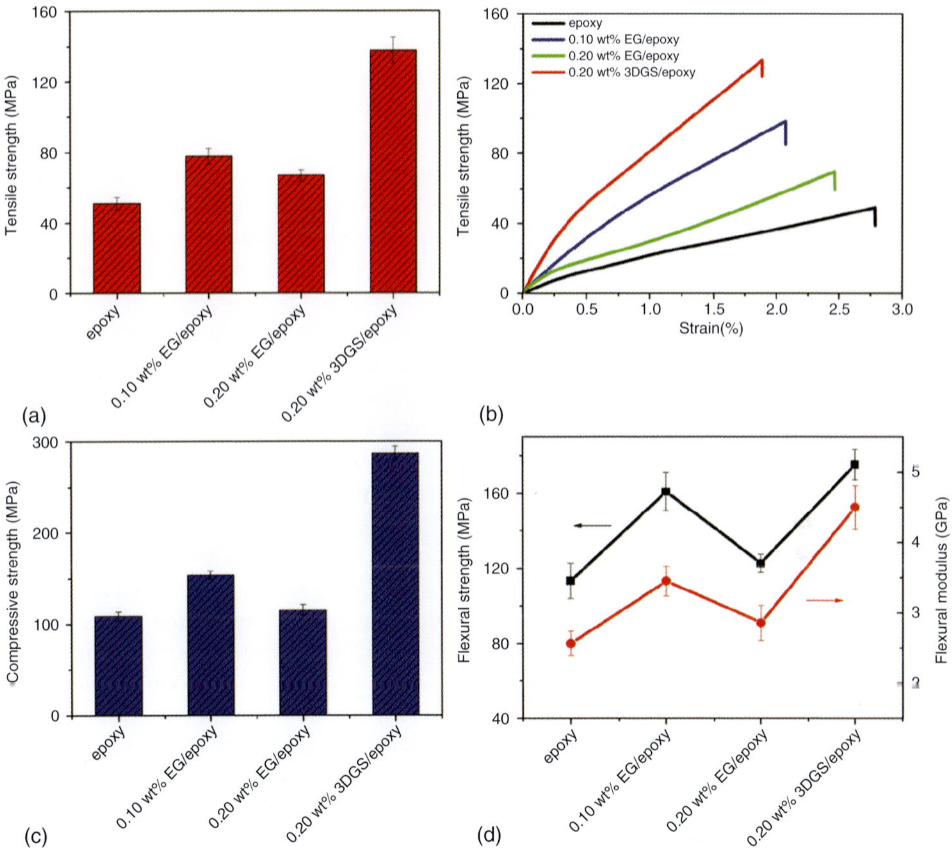

Figure 3.12 Mechanical properties of composites with different filler loadings: (a) tensile strength, (b) tensile stress–strain curves, (c) compressive strength, (d) flexural strength and modulus. (Reproduced from Ref. [134] with permission from the American Chemical Society.)

Jiang *et al.* modified GO by APTES functionalized silica nanoparticles (ATGO) and incorporated into the epoxy matrix [135]. They studied the effect of filler loadings on the tensile strength and impact strength both at RT and cryogenic temperature (CT). It is seen that the tensile strength and tensile modulus of the ATGO (1 wt%)/epoxy at CT were 29.2% and 22.0% higher than that of neat epoxy. The impact strength at RT was much higher than that at CT. It may be attributed to the presence of multifunctional groups and hence, good interaction between matrix and ATGO. Jiang *et al.* also modified the surface of silica nanoparticles by APTES and then attached with GO using isocyanate group-terminated flexible chains to synthesize modified filler (SATPGO) for the fabrication of epoxy composites [134]. The impact strength of the SATPGO filled (0.5 wt%) epoxy composite was found to be 154% and 92% higher than that of neat epoxy at RT and

CT, respectively. Tang *et al.* prepared RGO filled epoxy composites and investigated the influence of dispersion on the mechanical properties of the composites [136]. They reported improvement in KI_C values by 24% and 52% at 0.2 wt% RGO loadings. It indicates that the state and extent of dispersion greatly influence the end properties of composites. In a recent study, Jia *et al.* prepared interconnected porous structure graphene foam (GF)/epoxy composites and reported ~53% and 38% enhancements in flexural modulus and strength at ~0.2 wt% of GF [137].

The search for sustainable biodegradable polymers and the quest for mechanically robust and thermally enduring biodegradable polymer composites have stimulated many research activities. Gao *et al.* prepared poly GO/(propylene carbonate) (PPC) composites [138]. They reported 1000% increase in tensile strength, 1800% increase in Young's modulus, and 10 °C improvement in T_g. This encouraging result has shown new prospect of mechanically strong biodegradable polymers alternatives to petroleum-based polymers. Layek *et al.* prepared GO/sodium carboxymethylcellulose (NaCME) composite and reported ~188% and 154% increase in tensile strength and Young's modulus as compared to neat NaCME at only 1 wt% loading [139].

Zhang *et al.* reported the preparation of 3D graphene aerogel (GA)/polydimethylsiloxane (PDMS) composites (GAPC) by incorporating GA into PDMS by direct infiltration [140]. The well-aligned interconnected 3D GA frameworks/PDMS composites exhibited large mechanical deformability (compressive strain = 80% and tensile strain = 90%) along with stable piezo-resistance effect, and excellent electric Joule heating performance. With such outstanding properties, they may find applications in the field of stretchable electronic devices, ultra-large strain sensors, thermal interface materials, and so on.

3.6.2 Thermal Properties

Polymer composites with conductive fillers have emerged as one of the efficient heat dissipater materials. Despite interfacial resistance between the filler and polymer matrix, graphene has been extensively used to impart thermal conductivity to the host matrix due to its high thermal conductivity and flat surface area. Thermal expansion is also a very crucial property of graphene-based polymer composites. Due to negative CTE of graphene, high specific surface area and mechanical properties may reduce the CTE of the composites when it is deployed in polymers. The composites with high thermal conductivity and low CTE find potential applications as thermal interfacial material to reduce the thermal resistance and to dissipate heat in various electronic components. Li *et al.* prepared vertically aligned and densely packed multilayer graphene (MLG)/epoxy composite (AG/E) and reported maximum thermal conductivity of ~33.54 W m K^{-1} [141]. The thermal conductivity measured in perpendicular and parallel directions was found to be 16.75 and 33.54 W m K^{-1}, respectively. Further, the thermal conductivity of AG/epoxy composite exhibited a positive temperature response while increasing the temperature from 40 to 90 °C. Figure 3.13 represents the thermal conductivity

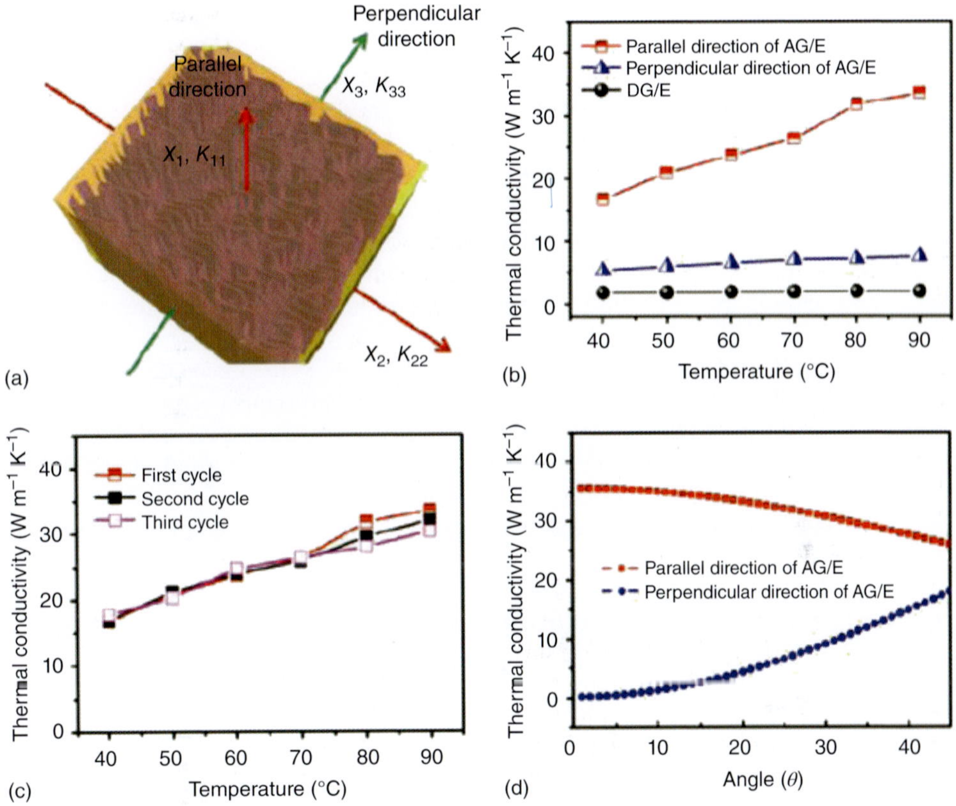

Figure 3.13 (a) Schematic diagram of parallel and perpendicular tensile strength, as well as electrical and thermal conductivities of AG/E in parallel and perpendicular directions. (b) Temperature dependence of thermal conductivities in AG/E and DG/E. (c) Thermal conductivity as a function of temperature for AG/E for three heating cycles. (d) The effect of the alignment extent on theoretical thermal conductivity in AG/epoxy. (Reproduced from Ref. [139] with permission from the American Chemical Society.)

of the MLG/epoxy composites along parallel and perpendicular directions. The commercially available GnPs/epoxy composites were prepared by applying high compression force to closure the gaps between the adjacent GnPs of large lateral dimensions. The thermal conductivity of the composites was found to be very high (12.4 W m K^{-1}) as compared to neat epoxy (0.2 W m K^{-1}) [142]. The addition of electrically insulating boron-nitride into the composites further improved the thermal conductivity and suppressed the electrical conductivity. Thermally conducting but electrically insulting composites so formed may find huge applications in the field of electronic capsulation.

Song *et al.* prepared noncovalently modified non-oxidized graphene flakes (GFs) by 1-pyrenebutyric acid (PBA) and the functionalized GFs (f-GFs) were incorporated into the epoxy matrix [143]. The thermal conductivity of the composites was found to be ~1.53 W m K^{-1} at 10 wt% of f-GF loadings. In addition to thermal conductivity, f-GFs also enhanced the mechanical properties of the epoxy composites and the modulus value was found to be 1.03 GPa at 1 wt% loading. Wang *et al.* incorporated GO into the epoxy matrix and investigated CTE behavior [144]. It has been found that the CTE values of the composites with 5 wt% GO loadings were decreased by 32% as compared to those of neat epoxy. Thermal conductivity of the composites was also increased fourfold as compared to that of neat epoxy. All these findings indicate that graphene/epoxy composites can be considered as promising candidates in the field of thermal management. Gallego *et al.* studied the effect of graphene platelets on the UV cured epoxy and reported the unprecedented 40 °C rise in T_g of the composites along with improved stiffness [145]. TEM results as shown in Figure 3.14 demonstrate the dispersion state of graphene platelets as well as the interaction between the graphene and matrix, which is reflected in the improvement of thermal and visco-elastic properties. Jia *et al.* prepared interconnected porous structure GF/epoxy and reported ~31 °C increase in T_g [137]. Araby *et al.* prepared electrically and thermally conductive GnPs/styrene butadiene rubber elastomeric composites by melt compounding and the reinforcing ability of GnPs was compared with that of CNTs [146]. It has been found that the percolation threshold in thermal conductivity was at 16.5 vol% GnPs loading. The highest thermal conductivity of the composites at 41.6 vol% GnP loading was ~240% higher as compared to that of neat epoxy. The significant improvement of thermal conductivity is encouraging for scaled-up production of elastomeric composites.

Design and control of interface is very crucial to achieve high-performance composites materials. Fang *et al.* tuned the interface between single-layer graphene nanosheets (SLGNs) and PS by controlling the grafting density and chain length of polymer through combination of diazonium addition and ATRP

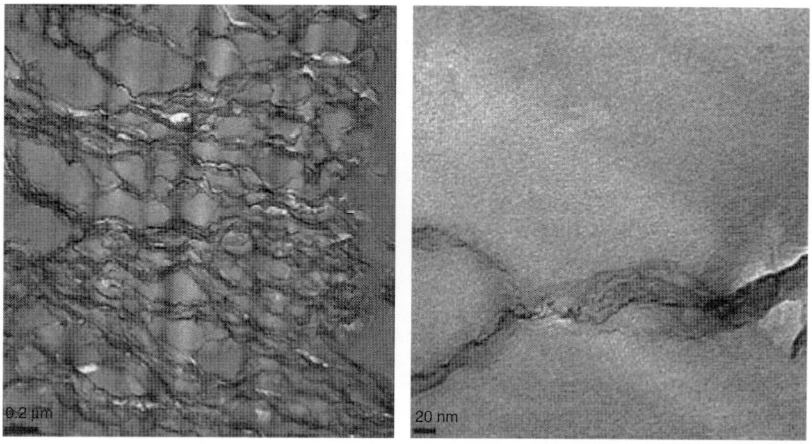

(a) (b)

Figure 3.14 TEM images of graphene-epoxy resin composite (1.5 wt% of filler loadings) at (a) low and (b) high magnification. (Reproduced from Ref. [145] with permission from the American Chemical Society.)

[145]. The T_g for the composites with high grafting density and low molecular weight polymer-grafting was found to be ~18 and 9 °C higher as compared to that of neat PS. Further, the thermal conductivity of the SLGN (2 wt%)/PS composites was ~2.6 times higher than that of neat PS. Patole *et al.* functionalized graphene sheets by PS nanoparticle following water-based in situ micro-emulsion polymerization and prepared PS composite film [147]. The composite exhibited ~16 °C rise in degradation temperature and temperature at the maximum reactive velocity obtained from differential thermal analysis peaks was increased by 15 °C at 20 wt%. The T_g was found to be increased by ~17 °C due to the good interfacial bonding that restricted the segmental motion of the polymer chains. Salavagione *et al.* prepared RGO/PVA composites by reducing GO in the presence of PVA matrix and coagulating the obtained material with 2-propanol [148]. The hydrogen-bonding interaction between PVA and RGO resulted in lowering the melting temperature (T_m) of the composites by ~35 °C and the crystallinity decreased to a half for the composites with 5 wt% of RGO loadings as compared to that of neat PVA. The T_g of the composite was increased by ~21 °C. The maximum degradation temperature of the composites was increased by more than 100 °C.

Wang *et al.* prepared GNS/PU composites by in situ polymerization [124]. The composite exhibited ~40 °C improvement in thermal stability with the addition of 2.0 wt% of GNS. The FG/PU composite exhibited ~50 °C improvement in thermal degradation temperature with 4 wt% of FG loadings [123]. These results revealed that graphene can reduce the diffusion of heat in the matrix and can act as a barrier to prevent the emission of thermally degraded small gaseous molecules. Myriad of studies have been carried out with graphene as filler material to fabricate polymer composites anticipating excellent material properties. However, apparently the intrinsic properties of graphene has not fully embedded in the composite. The limitation may lie in the processing and designing of the composite materials. If the graphene surface is meticulously designed in order to improve the interface with the polymer matrix, end materials with better properties can be expected. Liao *et al.* designed a flexible and covalently bonded interphase using two different molecular weight linear poly(oxyalkylene) amines with GO (D400-GO and D2000-GO) and incorporated into the PI matrix by in situ polymerization [122]. The T_g of the composite was increased by ~23.96 °C suggesting the restricted segmental motion of the polymer chains. Figure 3.15 shows the TMA curves of neat PI film and D400-GO/PI films.

Kuila *et al.* prepared graphene/PMMA composites by in situ emulsion polymerization followed by in situ reduction of GO and subsequent film casting in a glass petri dish [149]. The fine dispersion of graphene sheets in the PMMA matrix showed enhanced thermal stability of 26 °C as compared to that of neat PMMA.

Yun *et al.* studied the reinforcing effects of alkylated graphene oxide (AGO) on a nonpolar PP matrix and investigated the thermal and mechanical properties relative to 1D CNTs [150]. They reported ~33 °C increase in thermal degradation temperatures with the addition of only 1 wt% AGO into the PP matrix. Further, ~0.1 wt% of AGO increased the Young's modulus of the PP by more than 70%. The reinforcing ability of AGO relative to that of CNTs is mainly due to its 2-D

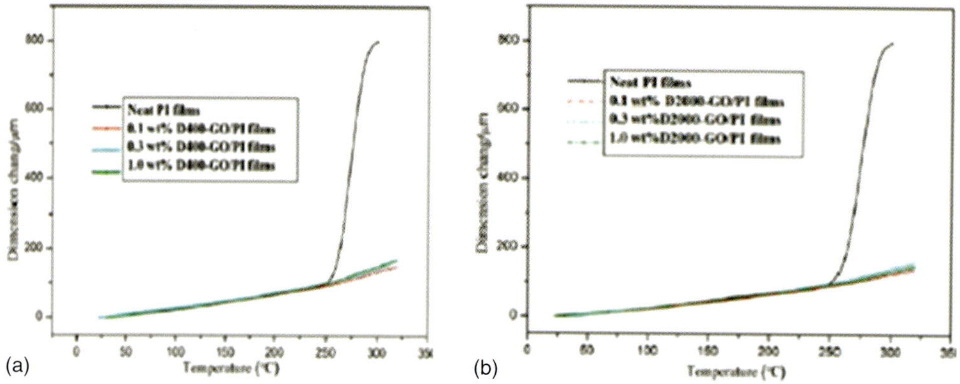

Figure 3.15 TMA curves of (a) neat PI film and D400-GO/PI films and (b) neat PI film and D2000-GO/PI films. (Reproduced from Ref. [122] with permission from the American Chemical Society.)

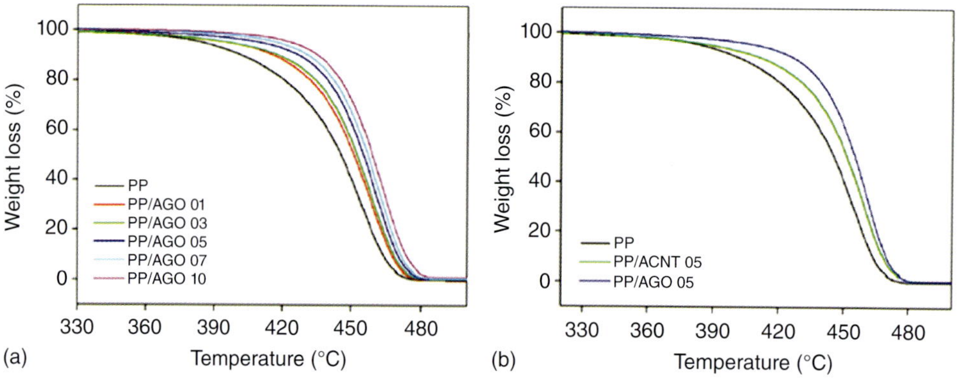

Figure 3.16 TGA profile of (a) alkylated graphene oxide/polypropylene composites and (b) different carbon fillers/polypropylene composites. (Reproduced from Ref. [148] with permission from Elsevier.)

structure and fine dispersion in the matrix, which may help to build strong interfacial strength. Figure 3.16 shows the TGA curve of AGO/PP composites and its comparison with CNTs. The above findings substantiate the fact that graphene and its derivatives are very useful to fabricate thermally conductive composites materials and also confirm that graphene sheets can act as a barrier to inhibit the propagation of heat in a polymer matrix, which results in improved thermal stability of polymer composites material.

3.6.3
Electrical Properties

The high aspect ratio and excellent electrical conductivity of graphene make it attractive as a filler to impart conductivity to the insulating polymer material.

Percolation threshold, ϕ_c, (the value at which conductive network of filler is established) of the graphene/polymer composites is anticipated to be very low due to its large intrinsic electrical conductivity and high surface area, thereby, enhancing the electrical properties of the composites at lower loading of graphene. Large numbers of polymers such as PMMA, PS, PU, epoxy PDMS, PVA, and so on have been used as matrix to incorporate graphene to fabricate conductive polymer composites. Conductive polymer composites find applications in electromagnetic shielding (EMI), antistatic coating and conductive paints. Rather than to use graphene prepared by CVD, thermally or chemically reduced graphene are used to impart conductivity to the composites mostly due to the processing simplicity and cost effectiveness. Stankovich *et al.* reported exfoliated phenyl isocyanate-treated GO/PS composite by solution mixing followed by in situ reduction and reported conductivity of 0.1 S cm^{-1} at ϕ_c of 0.1 volume fractions [151]. This unprecedented enhancement in electrical conductivity may be due to the large aspect ratio of CFG and its uniform distribution into the host matrix. Figure 3.17 shows the electrical conductivity of PS composites against filler loadings. Pham *et al.* fabricated chemically converted graphene (CCG)/PS composite by solution mixing followed by compression molding [152]. The authors revealed low percolation threshold of 0.19 vol% and high electrical conductivity of 72.18 S m^{-1} at loading of 2.45 vol%. The high electrical conductivity of CCG contributed to make conductive composite. Fine dispersion of CCG also made composites more stiff, which was reflected in increasing storage modulus value.

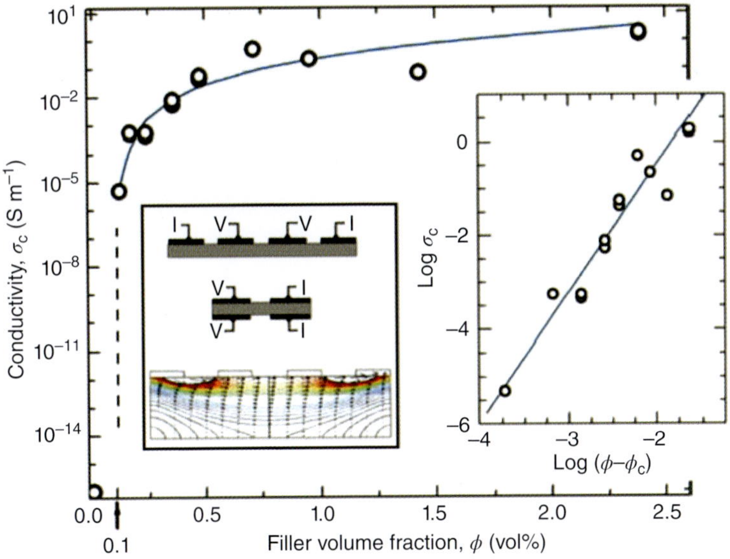

Figure 3.17 Electrical conductivity of isocyanate treated graphene/PS composites. Inset figure represents percolation threshold and set up for measurement. (Reproduced from Ref. [149] with permission from Nature Publication.)

Li *et al.* simultaneously functionalized and RGO with ODA and fabricated ODA-functionalized GO (GO-ODA)/PS composite [66]. The in situ reduction was confirmed by the enhanced electrical conductivity of GO-ODA and it was further increased during the compression-molding of the composites at 210 °C. The sharp transition in electrical conductivity (percolation threshold) for the composites was achieved at 0.45 vol% of GO-ODA loadings. The electrical conductivity of the composites at 0.46 and 0.92 vol% was found to be 9.2×10^{-4} and 4.6×10^{-1} S m^{-1}, respectively. Qi *et al.* compared the efficiency of graphene over CNT in improving electrical conductivity by fabricating PS composites. The graphene/PS composites showed 2–4 orders of magnitude higher electrical conductivity than CNT/PS composites [153]. The electrical conductivity of graphene/PS composites increased from $\sim 6.7 \times 10^{-12}$ to 349 S m^{-1} with the graphene loadings ranging from ~0.11 to 1.1 vol%. Further, the percolation threshold in electrical conductivity of the graphene/PS composites decreased approximately four- to fivefold while employing the exclusion (or selective localization) principle. Figure 3.18 shows the electrical conductivity of the graphene/PS and CNT/PS composites against filler content.

Wu *et al.* constructed 3D, compactly interconnected graphene network by self-assembling of PS and EVA polymer matrix [155]. The compact, continuous 3D architectures of graphene in polymer matrix facilitated electron flow seamlessly and exhibited electrical conductivity of 1083.3 and 260 S m^{-1} at 4.8 and 2.2 vol% graphene loadings, respectively. It is seen that the EVA composites not only exhibited high electrical conductivity but also showed enhanced flexibility. Self-assembly process could be proven to be advantageous as a compact structure can be designed with less contact resistance. In addition, this process is eco-friendly due to processing in an aqueous media. Figure 3.19 shows the electrical conductivity of PS composites and foam-like morphology of the 3D graphene

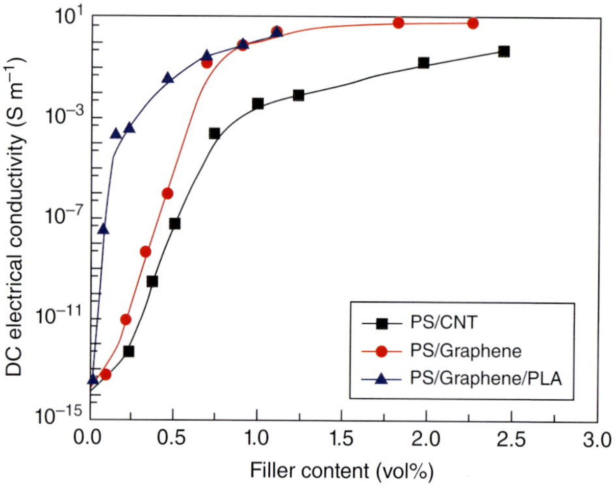

Figure 3.18 Variation of electrical conductivity of PS composite against filler loadings. (Reproduced from Ref. [154] with permission from the American Chemical Society.)

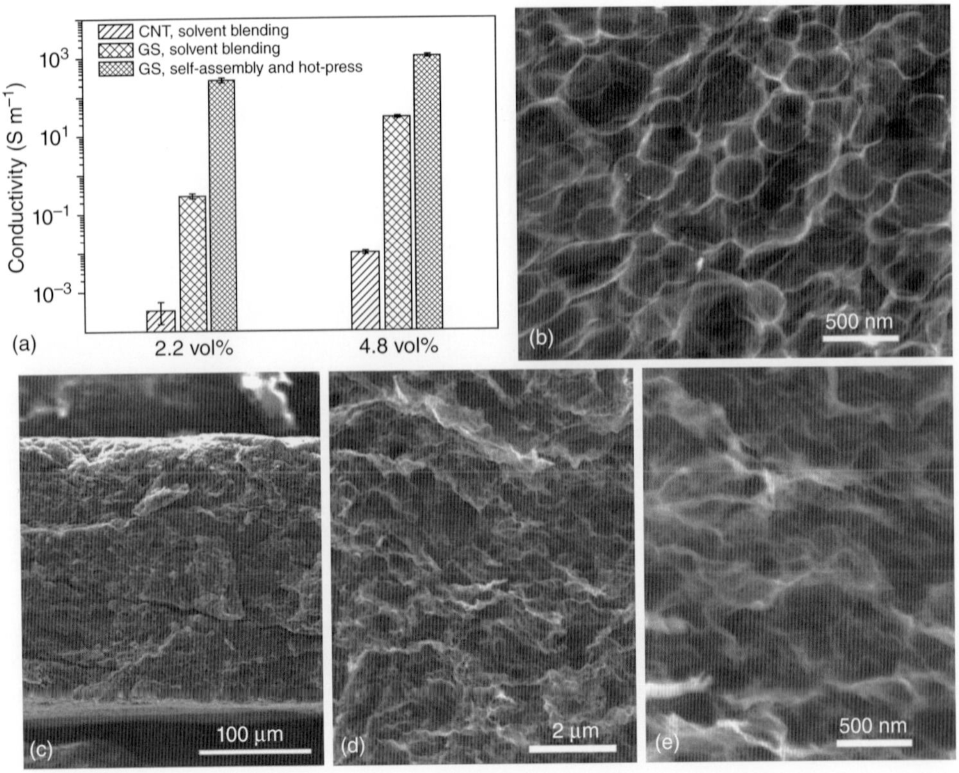

Figure 3.19 (a) Electrical conductivity comparison of the graphene and carbon nanotube–polystyrene composites fabricated by solution mixing and electrostatic self-assembly methods. (b) Top view and (c–e) cross-sectional SEM images of the remaining graphene skeleton after annealing the testing samples at high temperature under nitrogen atmosphere. (Reproduced from Ref. [151] with permission from John Wiley and Sons.)

after annealing at high temperature. Hsiao *et al.* prepared GO/WPU composites by LbL alternate deposition of cationic surfactant (didodecyldimethylammonium bromide, DDAB)-adsorbed positive charge GO and negative charge carrying GO suspension on the negatively charged WPU fibers followed by HI reduction to obtain RGO [156]. The RGO formed a network in the matrix and resulted in enhanced electrical conductivity of ~16.8 S m^{-1} for the composite.

Zhang *et al.* prepared graphene/PET composites by melt compounding technique with very low percolation threshold in electrical conductivity at 0.47 vol% [157]. The electrical conductivity of the composite was found to be 2.11 S m^{-1} at 3.0 vol% of graphene loadings. The high electrical conductivity that graphene imparted is due to its high aspect ratio, intrinsic electrical conductivity, and dispersion in the matrix. Vadukumpully *et al.* prepared surfactant-wrapped graphene nanoflakes/PVC ultrathin composite films. The percolation threshold in electrical conductivity of the composite was 0.6 vol% and showed a maximum electrical conductivity of 5.8 S m^{-1} at 6.47 vol% of graphene loading [115].

Shen *et al.* compared the efficiency of GO reduction of glucose and polyvinylpyrrolidone (PVP) and found glucose to be more effective than PVP [158]. The glucose-reduced GO (rGO-g)/PLA composites exhibited electrical conductivity of 2.2 S m^{-1} with 1.25 vol%. Furthermore, it is seen that glucose is a more efficient reducing agent as compared to that of the chemically RGO. Vuluga *et al.* prepared RGO/PMMA composites through biphasic process by radical addition pathway [159].The electrical conductivity of the composites was 3.4 S m^{-1}, which could be beneficial for EMI shielding purpose.

Recently, Jia *et al.* prepared an interconnected porous structure GF/epoxy composites by impregnation of epoxy resin into the 3D GF [137]. The composites showed the electrical conductivity of 3 S m^{-1} at the loading of 0.2 wt% GF. The significant increase in electrical conductivity may be due to the large surface area of GF and fast movement of charge carrier through the 3D network. The 3D GF is not only effective in enhancing electrical conductivity but also in influencing the mechanical and thermal properties significantly. Chen *et al.* synthesized 3D GF by template-directed CVD and fabricated GF/poly(dimethyl siloxane) composites [160]. The flexible interconnected network structure of GF acted as a fast transport channel for charge carrier and the composites showed electrical conductivity of ~1000 S m^{-1} even with loading of 0.5 wt%. The high electrical conductivity with low ϕ_C can be attributed to the unique interconnected network structure of GFs through which electrons can pass without interruption. Recently, Zhang *et al.* fabricated 3D GA–PDMS composites (GAPC). The interconnected 3D framework composites structure exhibited electrical conductivity of 114.7 S m^{-1} [140].

Zheng *et al.* performed one-step in situ reduction and polymerization process to fabricate polyamide RGO/PA-six composites using insulating GO and ε-caprolactam monomer [154]. Homogeneous dispersion of RGO in the PA six matrices resulted in electrical conductivity of ~0.028 S m^{-1} only at loading of ~1.64 vol% of GO. Yoonessi *et al.* prepared GNS/bisphenol A polycarbonate composite by both emulsion mixing and solution blending technique was subsequently compression molded at 287 °C [161]. The composites prepared by emulsion mixing and solution blending showed dc electrical percolation threshold of ~0.14 and 0.38 .vol% and the conductivity of 0.512 (emulsion mixing) and 0.226 S m^{-1} (solution blending) at 2.2 vol%. The improvement in electrical conductivity can be ascribed to the high aspect ratio, nanostructure directed assembly of GNS, and inherent conductivity of the GNS.

Recently, Vasileiou *et al* [162]. incorporated TRGO into the maleated LLDPE and its amino-pyridine derivatives by melt compounding [163]. The aromatic moieties on the backbone of polymer chains interacted with graphene surface through non-covalent interaction and formed electrically conducting graphene sheets. The percolation threshold in electrical conductivity for the non-covalently modified graphene was very low (0.5–0.9 vol%). Hyunwoo *et al.* coated ultrahigh molecular weight PE with GO sheets and then reduced using hydrazine subsequently resulting in the formation of segregated composites [164]. The two-step preparation technique might help to prevent aggregation of RGO sheets; the composites exhibited high electrical conductivity at a very low percolation threshold of

0.028 vol%. The process could be effective to fabricate electrically conductive composites with low percolation threshold. Kim *et al.* applied three most widely used procedures, that is, melt intercalation, solvent mixing, and in situ polymerization to fabricate thermoplastic PU composite using TRG and isocyanate treated GO as fillers [165]. Effective exfoliation of thin sheets made composite conductive at ~0.5 wt% of TRG loadings. It has been found that the solution mixing was better as compared to melt mixing or in situ polymerization in terms of distribution of fillers in the polymer matrix.

3.6.4
Dynamic Mechanical Properties

The dynamic mechanical properties such as storage modulus and loss angle tangent (tan δ) values depend on the interfacial bonding strength between the filler and matrix. Loss tangent depends on the restricted movement of polymer segments and is helpful for determining the T_g value. Roy *et al.* functionalized graphene with PEI and fabricated GO-PEI/nylon composites [127]. The PEI interacts with GO through hydrogen bonding as it contains several amine functional groups resulting in the homogeneous dispersion of PEI-GO into the nylon matrix. The storage modulus of the composite (0.25 wt%) was ~63% higher as compared to neat polymer and showed ~15 °C increment in thermal stability with ~0.35 wt% loadings.

Mittal *et al.* prepared TRGO filled HDPE, LLDPE, PP, PS, and PC composites by melt mixing technique [166]. The effects of filler content and type of polymer on the mechanical, thermal, and rheological properties have been investigated in detail. Among the studied composites, LLDPE composites showed 100% improvement in tensile modulus and 100% storage modulus with 7 wt% of TRGO loadings. These findings offered the explanation that not only polarity match between the matrix and filler but also the nature of polymer and processing conditions is important for improvement in physicochemical properties. Changsheng Xiang *et al.* reported GNS/PU composites by in situ polymerization [124]. Di-isocyanates were grafted onto the GNS followed by in situ polymerization. The composites exhibited improved tensile strength, storage modulus, electrical conductivity, and thermal stability due to the direct bond formation between GNS and PU, fine dispersion, and interfacial interaction. The storage modulus and the tensile strength were found to be increased by ~202% and ~239%, respectively, with the incorporation of 2.0 wt% of GNS. The effect of RGO and surface-modified RGO on the mechanical properties of PI composite was investigated in detail [167]. The surface-modified RGO filled composites showed 25–30% higher storage modulus than that of RGO filled composite. The improvement in mechanical properties can be attributed to the fine dispersion and the stiff benzene ring of PI moieties. Figure 3.20 shows the variation of storage modulus for RGO and surface treated RGO filled PI composites at different temperature. It shows that the storage modulus increased with increasing RGO or surface treated RGO content.

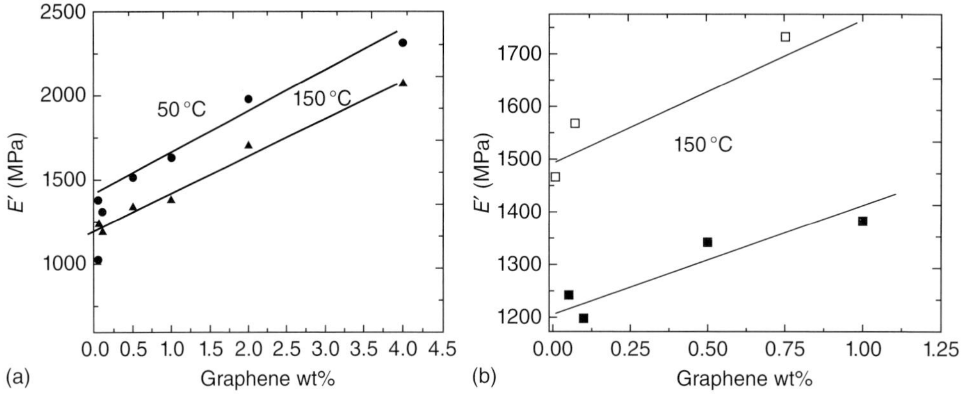

Figure 3.20 (a) Dynamic tensile moduli of graphene polyimide nanocomposites at 50 (b) and 150 °C (2) as a function of graphene content. (b) Dynamic storage moduli of the polyimide nanocomposites containing graphene (9) and rigid imidized graphene (0) at 150 °C. (Reproduced from Ref. [166] with permission from American Chemical Society.)

Layek *et al.* prepared GO/sodium carboxymethyl cellulose (NaCME) composite by solution casting method [130]. Hydrogen bond between the hydrophilic group of GO and NaCME resulted molecular level dispersion of GO in NaCME, which is reflected in the 174% increase in storage modulus at the loading of 1 wt% GO along with the increase 188% and 154% in tensile stress and Young's modulus respectively. Figure 3.21 shows the schematic for the hydrogen bonding interaction between GO and NaCME. In another report of biodegradable polymer, poly (propylene carbonate) (PCC) an increase of 1600 MPa was recorded with the addition of 1 wt% GO along with 10 °C rise in T_g [138].

Tang *et al.* dispersed RGO sheets into the epoxy matrix with and without ball mill mixing [41]. The effect of dispersion state of graphene on the various properties of the composites have been investigated in detail. It is seen that the well-dispersed RGO significantly improved the T_g, tensile and flexural strength of the composites as compared to the poorly dispersed RGO. The increase in T_g was ~11 °C with the addition of only 0.2 wt% well-dispersed RGO to epoxy. Figure 3.22 shows the dynamical mechanical properties of well-dispersed and poorly dispersed RGO/epoxy composites and it is apparent that storage modulus improvement is comparatively high in the case of highly dispersed RGO. Therefore, it is seen that the fine dispersion of graphene nanoparticles is very crucial to property improvements.

3.7
Application of Graphene Based Polymer Composites

Graphene/polymer composites have been studied for a myriad of applications such as engineering materials, gas barrier, sensor, EMI shielding, fire retardant

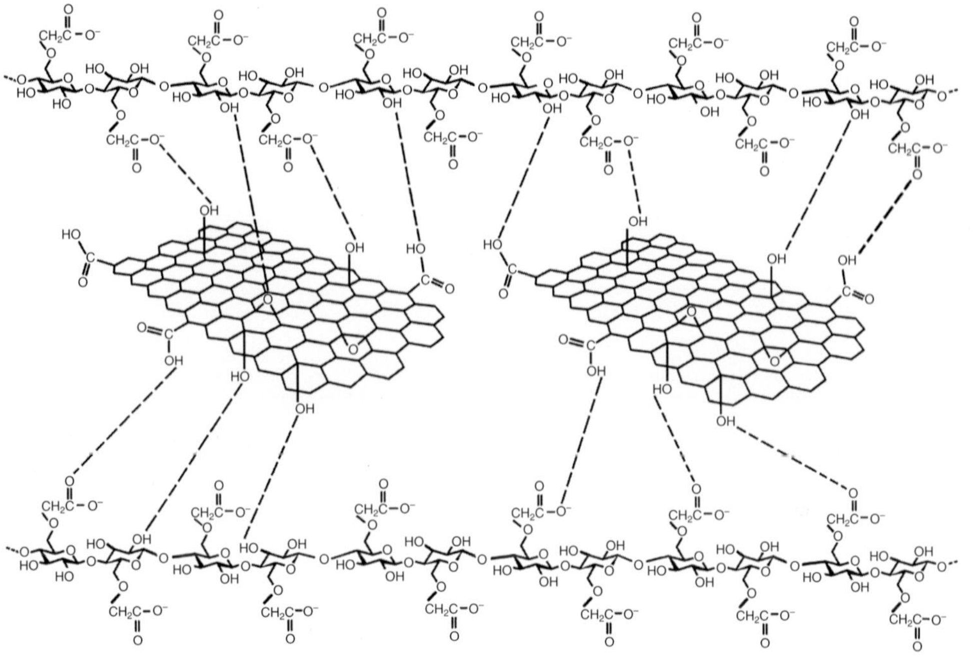

Figure 3.21 Hydrogen bonding interaction between GO and NaCME. (Reproduced from Ref. [130] with permission from John Wiley and Sons.)

materials, turbine blades, conductive composite, in automotive, aircraft, and marine industries, and so on [25, 96]. Advanced polymer composites materials are becoming very crucial components in the design of new windmill blades, aircraft components, and other various applications, which require light-weight, high-strength materials. Graphene-based polymer composites are promising materials in the pursuit of developing structural components with a high strength-to-weight ratio. The amount of filler loading, dispersion of graphene in the matrix, and interaction between the components are very crucial parameters to determine the prospect of graphene-based composites for real-world applications.

3.7.1
Gas Barrier

The large surface area and the non-permeable flat sheet-like structure of graphene has allowed its applications in the areas of gas barrier applications. If graphene or its derivatives are finely dispersed in polymer matrix, the "tortuous path" so generated by graphene would inhibit the molecular diffusion of gaseous molecules to remarkably reduce permeability. It is imperative to design thin film polymeric materials that can prevent diffusion of moisture, which is desirable in the field of food packing, electronics, and fuel cells. Yang *et al.* reported thin film structure

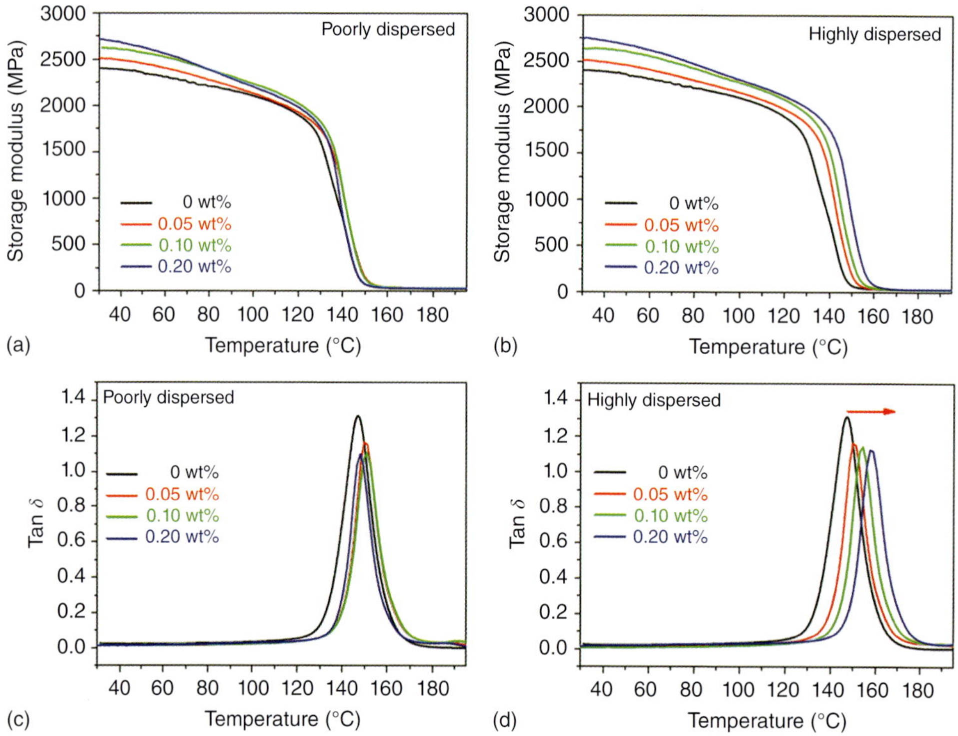

Figure 3.22 Dynamic mechanical properties of cured epoxy composites containing: (a) and (c) poorly dispersed RGO; (b) and (d) highly dispersed RGO. (Reproduced from Ref. [41] with permission from Elsevier.)

of GO/PEI composite prepared by LbL assembly of alternate GO and PEI layers to investigate the oxygen barrier properties of these composites [168]. The significant reduction of oxygen as well as carbon dioxide permeation properties of the composites has the potential to prevent hydrogen permeation. The remarkable performance of the composites with regard to barrier properties can be ascribed to the "tortuous path" that the packed nano brick structure composites created prevented the diffusion of gases. Liu *et al.* prepared RGO/PEI composite film, where PEI performed ternary roles of a reductant, modifier, and polymer host [169]. The brick and mortar type structure so formed by vacuum filtration of PEI-RGO showed gas barrier properties and the properties increased with increasing PEI content whereas conductivity got decreased. The gas barrier properties can be attributed to the finely wrapping of graphene sheet by polymer chains. The FG/nylon composites has shown very efficient water vapor and gas barrier properties at very small loading of FG. The composites showed decreased water vapor and oxygen permeability by 49% and 47%, respectively, with 0.3 wt% of FG loadings [126]. It can be ascribed to the intrinsic flexibility of graphene, 2D structure, high aspect ratio, and fine dispersion of FG. Layek *et al.* showed that the oxygen permeability coefficient of the SPG/PVDF composites was significantly lower than that of the neat PVDF [85]. Based on the above results, graphene/polymer composite can be used as food packaging materials. Specific examples of applications include packaging for cheese, cereals, edible oils, confectionery, fruit juice, dairy products, processed meats, beer, carbonated drinks, and so on. The use of graphene/polymer composite as packaging materials would be expected to significantly improve the shelf life of many types of food. The transportation and storage of hydrogen gas are highly dangerous due to its continuous diffusion through the wall of the container. Therefore, graphene filled polymer composites can be used as coating materials to prevent the diffusion of hydrogen gas.

3.7.2 Sensor

Precisely designed thin films (sensing layer) of graphene/polymer composites have drawn significant interest in the paradigm of sensing materials. Conducting polymer/graphene composites are important as sensing materials due to its effectiveness as senors, low cost of production, and high electrical response. The excellent electrical conductivity of graphene is the main factor that makes it suitable for use in sensing technology. Wu *et al.* fabricated graphene/PANI composites by chemical oxidative polymerization for the detection of NH_3 [170]. It is seen that the sensitivity of graphene/PANI composites was higher than that of neat PANI with the detection limit of 1 ppm. Yang *et al.* prepared CVD graphene/PMMA composites for the development of flexible sensors on paper substrate that can detect NO_2 at the concentration of 200 ppm under different strains [171]. Bai *et al.* tested the sensing ability of graphene/polypyrrole hydrogels against NH_3 and found good sensing ability toward NH_3 [172]. It has been found that the dG-O/PVA composite film can be used as moisture sensors [117].

3.7.3 EMI Shielding

Graphene-based thermoplastic/thermoset composite has attracted significant attention as EMI shielding materials due to the low percolation threshold in the mechanical and electrical properties of the graphene filled composites. It is very essential to protect interference prone equipment from malfunctioning. EMI not only hampers electronic devices but is also critical to human health. Thus, metal and metallic composites were the choice for shielding purposes due to their high electrical conductivity. However, difficulty in processing, heavy weight, and its chemical instability has urged to look beyond it. Therefore, graphene filled polymer composites has started getting attention as EMI shielding materials due to its light weight, large specific surface area and good electrical conductivity.

Yan *et al.* fabricated RGO/PS composites via high pressure solid-phase compression molding for EMI shielding applications [173]. Multi-facet segregated architecture of RGO in PS matrix provides many interfaces that facilitated efficient shielding of EMI. The shielding effectiveness of the composite was found to be 45.1 dB at a loading level of only 3.47 vol%. Further, the compressive strength and compressive modulus of the composite was 94% and 40% higher as compared to those of neat PS. Figure 3.23 shows the variation of electrical conductivity, EMI shielding effectiveness, and mechanical properties. The electrical conductivity of f-G/PVDF composite was 10^{-4} S m^{-1} for 0.5 wt% of f-G loadings. Further, the EMI shielding effectiveness of the composite was found to be ~20 dB in X-band (8–12 GHz) region and 18 dB in broadband (1–8 GHz) region for 5 wt% of f-G loadings [174]. The EMI shielding efficiency of 21 dB over the frequency range of 8.2–12.4 GHz (X-band) was recorded for the 0.52 vol% of RGO filled epoxy composite [106]. Hsiao *et al.* showed the EMI SH effectiveness of ~34 dB over the frequency range of 8.2–12.4 GHz for the RGO/PU composite [156].

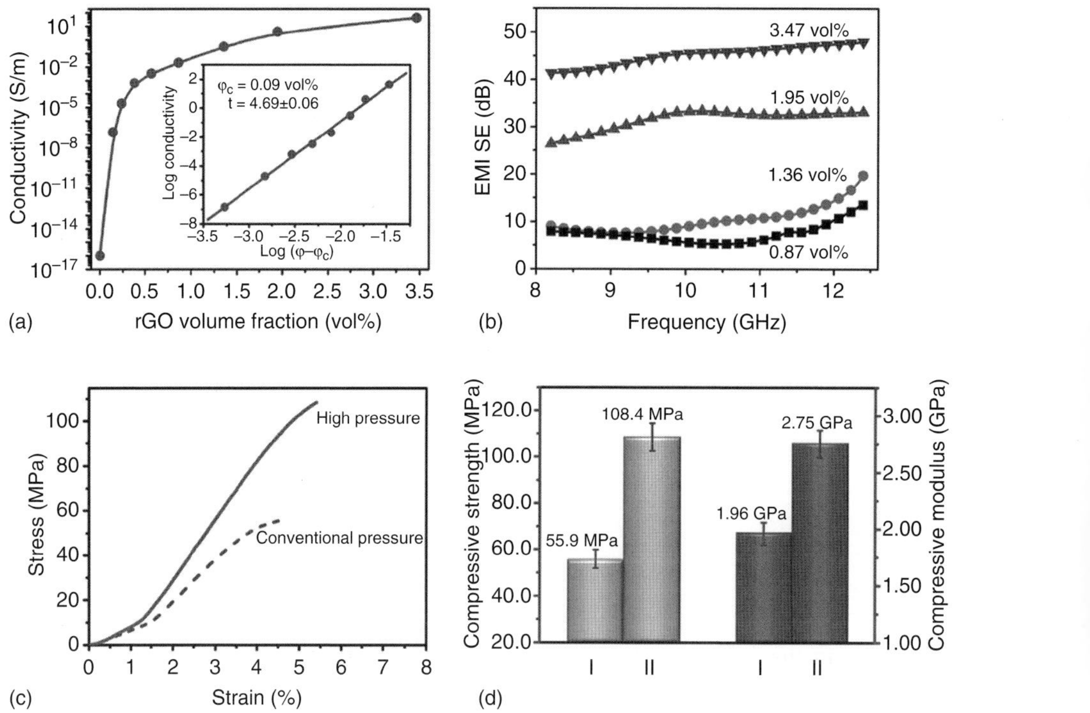

Figure 3.23 (a) Electrical conductivity versus rGO loading for the s-rGO/PS composites; (b) EMI SE as a function of frequency in X-band range for the composites; (c) typical stress–strain curves of the s-rGO/PS composites with 1.95 vol% rGO molded under high and conventional pressure; and (d) compressive strength and modulus of the s-rGO/PS composites. I and II indicate composites molded under conventional and high pressure, respectively. (Reproduced from Ref. [173] with permission from John Wiley and Sons.)

3.7.4 Flammability Reduction

Polymers are prone to burn easily, so, it has become crucial to make them fire resistive as the horizon for polymer composite applications is increasing day by day. One of the adopted strategies to improve the fire safety of polymers is to incorporate fire retardant additives. Being two-dimensional with high specific surface area and high thermal stability, graphene has the potential to be a fire retardant. If the interfacial interaction between graphene and the polymer matrix is good then it can act as barrier material that can prevent the release of heat and degrade small gaseous molecules efficiently.

Wang *et al.* used octa-aminophenyl polyhedral oligomericsilsesquioxanes (OapPOSS) for the simultaneous functionalization and reduction of GO. The OapPOSS-rGO was incorporated into the epoxy matrix to study the thermal stability and flammability of the composites [175]. The peak heat release rate, total heat release, and CO production rate of OapPOSS-rGO/epoxy composites was found to be 49%, 37%, and 58%, respectively. Furthermore, the onset degradation temperature was improved by 43 °C. The significant improvement in such properties can be ascribed to the fact that OapPOSS-rGO forms insulating charred layer that may act as a barrier and prevent the underlying polymer from burning. Bao *et al.* functionalized GO with char-catalyzing agents hexachlorocyclotriphosphazene (HCCP) and hydroxyethyl acrylate (HEA) for the development of fire retardant graphene/PS composites [176]. The physical barrier effect of FGO, its char formation ability, and polymer-FGO adhesion contributed to the improvement of fire safety of the composites. The composites exhibited improved peak CO release rate (66% decreases), peak CO_2 release rate (54% decrease), peak heat release rate (53% decrease), thermal degradation rate (30% decrease), and total heat release (38% decrease).

3.7.5 Automotive and Aircrafts

Graphene-based polymer composites are promising materials that could be used to make lighter, more fuel-efficient aircraft and automobile parts. With the first breakthrough in 1991 by Toyota Motor Co. significant research activities and development have occurred focusing polymer composites as promising candidates in the field of automotive sector. Polymer composites are promising materials that exhibit a plethora of properties such as superb mechanical properties, high modulus and dimensional stability, excellent thermal properties, enhanced scratch and mar resistance, flame retardant properties, high impact resistance, and barrier properties, owing to which they can be perceived as an alternative to metals in automotive and other applications [177]. The factors that stimulated the used of polymer composites as automotive components are reduction in vehicle's weight that improves engine efficiency, reduction in CO_2 emission and better performance in terms of safety and comfort. As graphene has excellent properties and is the stiffest material if dispersed seamlessly in polymer matrix, composites so formed can be used to develop strong and lightweight automotive components. The University of Sunderland is leading a project with five research partners from Italy, Spain, and Germany through Graphene Flagship Competitive Call to develop novel graphene-based polymer composites material and their application in the development of lighter, greener, and improved crash-worthiness vehicle. The research done on GNP and CNT reinforced technology by UK based Haydale and The School of Engineering at Cardiff University have reported a 13% increase in compression strength and a 50% increase in compression after impact performance, which is crucial in high-performance composites such as aircraft wings [178]. Another research group dispersed GNP at a loading of 15 wt% in material matrix, which resulted in electrical conductivity of $1000\,S\,m^{-1}$ and which may find use in aircraft brake and automotive friction plate components [179].

3.7.6 Turbine Blades

Lately, wind power has attracted attention as one of the emerging sources of energy [180]. Those countries having widespread coastlines and high onshore wind activity are intensively pursuing new ways in the direction of harnessing wind energy and converting it to useful energy as this is pollution free, requires minimum raw materials and low cost of power generation [181]. Development of components such as wind blades in wind power technologies becomes practical only when the cost of production of material is minimum and material is strong/lightweight that consumes less energy to function. It has been found that graphene is more superior to CNTs in reinforcing the composites' properties and may produce lightweight stiffer components than CNTs. For example, graphene-incorporated epoxy composites can be used to produce windmill blades or aircraft machineries. This will make the pursuit of lightweight, strong, and longer windmill blades a necessity, which will increase the generation of electricity from blades.

3.7.7 Others

Graphene/polymer composites can be used to construct various submarine structures such as bow domes on combatant submarines, periscope fairings on nuclear submarines, pressure hulls, and specialized sonar transducers. The composite materials can also be used in the light weight speed boat, life vest, swimming pool kits, and so on. The other possible applications of graphene/polymer composite materials include the manufacture of children entertainment equipments, highly durable industrial and domestic plastic products, scratch/abrasion resistant materials, anti corrosion paints, and so on.

3.8
Conclusions and Outlook

This chapter deals with graphene and its derivatives integrated polymer composites. Preparation and properties of graphene derivatives have been summarized and various techniques of incorporating graphene into the matrix have been discussed with appropriate examples. Various techniques of composite fabrication and their advantages or disadvantages have also been highlighted and critical issues relating to dispersion, distribution of graphene layers are discussed in detail. It is seen that the selection of proper dispersion technique, suitable tuning of graphene surface with strong interface and alignment of layers are critical to achieve robust, conductive composites. The graphene composites display improvement in mechanical, electrical, thermal, barrier, fire retardant properties. It can be noted that effective surface modification of graphene facilitates the fabrication of high performance composites. It is noteworthy that small amounts of graphene is sufficient to improve the physicochemical properties of graphene. Graphene/polymer composites find application in the fields of EMI shielding, automotive and marine industry, aircraft, flame retardant materials, and so on.

The major challenge that prevails in the area of graphene/polymer composites is the control of structure as well as mechanism of load transfer across the interface of graphene polymer matrix. If the structure of the composites could be controlled then it would be a breakthrough in the direction of understanding the underlying mechanism and process control. More effort should be made in the direction of advanced stimulation technique to understand the structure/property relationship of existing materials and in the development of novel materials with enhanced properties. Apart from structure control, effective uniform dispersion and distribution of graphene sheets are also challenging tasks. Therefore, new techniques of fabrication should be envisaged, through which the structure and distribution of graphene sheets can be controlled to develop multifunctional material as expected from a superb material such as graphene. Techniques such as LbL, where polymer and graphene are alternately deposited could be one probable measure to maximize dispersion and distribution, as well as the structure control of graphene nanosheets. However, this technique is able to generate highly ordered layered composites films but mass production of bulk items is difficult.

Improvement of interaction between graphene and polymer matrix is also very crucial to transfer the properties of graphene. Surface tuning is very important to construct a strong interface, which eventually determines the stress transfer. As pristine graphene does not hold vulnerable functionalities for further modification only non-covalent interactions are possible, which is not much effective in creating a strong interface, and so, GO and derivatives are considered as promising candidates for reinforcement. The surface of GO can be tuned by way of different interactions, which can be helpful to design strong interfaces. Polarity match between the graphene derivatives and matrix may result in better performance and multifunctional surfactants may help to generate a strong interface through various interactions.

References

1. Landsiedel, R., Hock, L.M., Kroll, A., Hahn, D., Schnekenburger, J., Wiench, K., and Wohlleben, W. (2010) Testing metal-oxide nanomaterials for human safety. *Adv. Mater.*, **22**, 1–27.
2. Saheb, D.N. and Jog, J.P. (1999) Natural fiber polymer composites: a review. *Adv. Polymer Technol.*, **18**, 351–363.
3. Balazs, A.C., Emrick, T., and Russell, T.P. (2006) Nanoparticle polymer composites: where two small worlds meet. *Science*, **314**, 1107–1110.
4. Ahmed, R.M. and Bashir, S.M.E. (2010) Structure and physical properties of polymer composite films doped with fullerene nanoparticles. *Int. J. Photoenergy*, **2011**, 1–6.
5. Fan, J., Shi, Z., Tian, M., Wang, J., and Yin, J. (2012) Unzipped multiwalled carbon nanotube oxide/multiwalled carbon nanotube hybrids for polymer reinforcement. *ACS Appl. Mater. Interfaces*, **4**, 5956–5965.
6. Le Baron, P.C., Wang, Z., and Pinnavaia, T.J. (1999) Polymer-layered silicate nanocomposites: an overview. *Appl. Clay Sci.*, **15**, 11–29.
7. Wang, M., Braun, H.G., and Meyer, E. (2002) Patterning of polymeric/inorganic nanocomposite and nanoparticle layers. *Chem. Mater.*, **14**, 4812–4818.
8. Carter L. W., Hendricks J. G., and Bolley D. S. (1950) Elastomer Reinforced with a Modified Clay. US Patent 2531396, National Lead Company.
9. Usuki, A., Kojima, Y., Kawasumi, M., Okada, A., Fukushima, Y., and Kurauchi, T. (1993) Synthesis of nylon 6-clay hybrid. *J. Mater. Res.*, **8**, 1179–1184.
10. Hu, K., Kulkarni, D.D., Choi, I., and Tsukruk, V.V. (2014) Graphene–polymer nanocomposites for structural and functional applications. *Prog. Polym. Sci.*, **39**, 1934–1972.
11. Mallick, P.K. (1993) *Fiber-reinforced Composites: Materials, Manufacturing and Design*, Marcel Dekker, Inc., New York, p. 584.
12. Kuila, T., Srivastava, S.K., and Bhowmick, A.K. (2009) Rubber/LDH nanocomposites by solution blending. *J. Appl. Polym. Sci.*, **111**, 635–641.
13. Yasmin, A., Luo, J.J., and Daniel, I.M. (2006) Processing of expanded graphite reinforced polymer nanocomposites. *Compos. Sci. Technol.*, **66**, 1182–1189.
14. Cheng, Q., Wang, B., Zhang, C., and Liang, Z. (2010) Functionalized carbon-nanotube sheet/bismaleimide nanocomposites: mechanical and electrical performance beyond carbon-fiber composites. *Small*, **6**, 763–767.
15. Choia, Y.K., Gotoh, Y., Sugimotoa, K., Songa, S.M., Yanagisawac, T., and Endo, M. (2005) Processing and characterization of epoxy nanocomposites reinforced by cup-stacked carbon nanotubes. *Polymer*, **46**, 11489–11498.
16. Kim, B.C., Park, S.W., and Lee, D.G. (2008) Fracture toughness of the nanoparticle reinforced epoxy composite. *Compos. Struct.*, **86**, 69–77.
17. Mamedov, A.A. and Kotov, N.A. (2000) Free-standing layer-by-layer assembled films of magnetite nanoparticles. *Langmuir*, **16**, 5530–5533.
18. Zheng, W., Lu, X., and Wong, S.C. (2004) Electrical and mechanical properties of expanded graphite-reinforced high-density polyethylene. *J. Appl. Polym. Sci.*, **91**, 2781–2788.
19. Wang, D.H., Sihn, S., Roy, A.K., Baek, J., and Tan, L.S. (2010) Nanocomposites based on vapour-grown carbon nanofibers and an epoxy: functionalization, preparation and characterization. *Eur. Polym. J.*, **46**, 1404–1416.
20. Mazinani, S., Ajji, A., and Dubois, C. (2009) Morphology, structure and properties of conductive PS/CNT nanocomposites electrospun mat. *Polymer*, **50**, 3329–3342.
21. Renukappa, N.M., Siddaramaiah , Sudhaker Samuel, R.D., Sundara Rajan, J., and Lee, J.H. (2009) Dielectric properties of carbon black: SBR composites. *J. Mater. Sci. Mater. Electron.*, **20**, 648–656.
22. Liu, Y., Wang, A., and Claus, R. (1997) Molecular self-assembly of TiO_2/polymer nanocomposite films. *J. Phys. Chem. B*, **101**, 1385–1388.

23. Potts, J.R., Dreyer, D.R., Bielawski, C.W., and Ruoff, R.S. (2011) Graphene-based polymer nanocomposites. *Polymer*, **52**, 5–25.
24. Wang, J., Cheng, Q., Lin, L., Chen, L., and Jiang, L. (2013) Understanding the relationship of performance with nanofiller content in the biomimetic layered nanocomposites. *Nanoscale*, **5**, 6356–6362.
25. Kuilla, T., Bhadra, S., Yao, D., Kim, N.H., Bose, S., and Lee, J.H. (2010) Recent advances in graphene based polymer composites. *Prog. Polym. Sci.*, **35**, 1350–1375.
26. Chen, G.H., Wu, D., Weng, W.G., and Wu, C.L. (2003) Exfoliation of graphite flake and its nanocomposites. *Carbon*, **41**, 619–621.
27. Liao, S.H., Yen, C.Y., Weng, C.C., Lin, Y.F., Ma, C.C.M., and Yang, C.H. (2008) Preparation and properties of carbon nanotube/polypropylene nanocomposite bipolar plates for polymer electrolyte membrane fuel cells. *J. Power Sources*, **185**, 1225–1232.
28. Li, Q., Siddaramaiah , Kim, N.H., Yoo, G.H., and Lee, J.H. (2009) Positive temperature coefficient characteristic and structure of graphite nanofibers reinforced high-density polyethylene/carbon black nanocomposites. *Composites Part B*, **40**, 218–224.
29. Chrissafis, K., Paraskevopoulos, K.M., Stavrev, S.Y., Docoslis, A., Vassiliou, A., and Bikiaris, D.N. (2007) Characterization and thermal degradation mechanism of isotactic polypropylene/carbon black nanocomposites. *Thermochim. Acta*, **465**, 6–17.
30. Sharmila, T.K.B., Nair, A.B., Abraham, B.T., Beegum, P.M.S., and Thachil, E.T. (2014) Microwave exfoliated reduced graphene oxide epoxy nanocomposites for high performance applications. *Polymer*, **55**, 3614–3627.
31. Novoselov, K.S., Geim, A.K., Morozov, S.V., Jiang, D., Zhang, Y., Dubonos, S.V., Grigorieva, I.V., and Firsov, A.A. (2004) Electric field effect in atomically thin carbon films. *Science*, **306**, 666–669.
32. Stankovich, S., Dikin, D.A., Dommett, G.H.B., Kohlhaas, K.M., Zimney, E.J., and Stach, E.A. (2006) Graphene-based composite materials. *Nature*, **442**, 282–286.
33. Lee, C., Wei, X.D., Kysar, J.W., and Hone, J. (2008) Measurement of the elastic properties and intrinsic strength of monolayer graphene. *Science*, **321**, 385–388.
34. Stoller, M.D., Park, S.J., Zhu, Y.W., An, J.H., and Ruoff, R.S. (2008) Graphene-based ultracapacitors. *Nano Lett.*, **8**, 3498–3502.
35. Nair, R.R., Blake, P., Grigorenko, A.N., Novoselov, K.S., Booth, T.J., Stauber, T., Peres, N.M.R., and Geim, A.K. (2008) Fine structure constant defines visual transparency of graphene. *Science*, **320**, 1308.
36. Balandin, A.A., Ghosh, S., Bao, W., Calizo, I., Teweldebrhan, D., Miao, F., and Lau, C.N. (2008) Superior thermal conductivity of single-layer graphene. *Nano Lett.*, **8**, 902–907.
37. Ramanathan, T., Abdala, A.A., Stankovich, S., Dikin, D.A., Alonso, M.H., and Piner, R.D. (2008) Functionalized graphene sheets for polymer nanocomposites. *Nat. Nanotechnol.*, **3**, 327–331.
38. Lee, Y.R., Raghu, A.V., Jeong, H.M., and Kim, B.K. (2009) Properties of waterborne polyurethane/functionalized graphene sheet nanocomposites prepared by an in situ method. *Macromol. Chem. Phys.*, **210**, 1247–1254.
39. Xu, Y., Wang, Y., Jiajie, L., Huang, Y., Ma, Y., and Wan, X. (2009) A hybrid material of graphene and poly (3,4-ethyldioxythiophene) with high conductivity, flexibility, and transparency. *Nano Res.*, **2**, 343–348.
40. Quan, H., Zhang, B., Zhao, Q., Yuen, R.K.K., and Li, R.K.Y. (2009) Facile preparation and thermal degradation studies of graphite nanoplatelets (GNPs) filled thermoplastic polyurethane (TPU) nanocomposites. *Composites Part A*, **40**, 1506–1513.
41. Tang, L.C., Wan, Y.J., Yan, D., Pei, Y.B., Zhao, L., Li, Y.B., Wu, L.B., Jiang, J.X., and Lai, G.Q. (2013) The effect of graphene dispersion on the mechanical properties of graphene/epoxy composites. *Carbon*, **60**, 16–27.

42. Salavagione, H.J., Martínez, G., and Ellis, G. (2011) Physics and applications of graphene – experiments. *Nanotechnol. Nanomater.*, 169–192, edited by Sergey Mikhailov, ISBN 978-953-307-217-3.
43. Eizenberg, M. and Blakely, J.M. (1979) Carbon monolayer phase condensation on Ni(III). *Surf. Sci.*, **82**, 228–236.
44. Aizawa, T., Souda, R., Otani, S., Ishizawa, Y., and Oshima, C. (1990) Anomalous bond of monolayer graphite on transition metal carbide surfaces. *Phys. Rev. Lett.*, **64**, 768–771.
45. Lu, X., Yu, M., Huang, H., and Rouff, R.S. (1990) Tailoring graphite with the goal of achieving single sheets. *Nanotechnology*, **10**, 269–272.
46. Keidar, M., Shashurin, A., Volotskova, O., Raitses, Y., and Beilis, I.I. (2010) Mechanism of carbon nanostructure synthesis in arc plasma. *Phys. Plasmas*, **17**, 057101–0571019.
47. Berger, C., Song, Z., Li, X., Wu, X., Brown, N., and Naud, C. (2006) Electronic confinement and coherence in patterned epitaxial graphene. *Science*, **312**, 1191–1196.
48. Li, D., Muller, M.B., Gilje, S., Kaner, R.B., and Wallac, G.G. (2007) Process able aqueous dispersions of graphene nanosheets. *Nat. Nanotechnol.*, **3**, 101–105.
49. Kim, H., Abdala, A.A., and Macosko, C.W. (2010) Graphene/polymer nanocomposites. *Macromolecules*, **43**, 6515–6530.
50. Hummers, W.S. and Offeman, R.E. (1958) Preparation of graphitic oxide. *J. Am. Chem. Soc.*, **80**, 1339–1339.
51. Stankovich, S., Dikin, D.A., Piner, R.D., Kohlhaas, K.A., Kleinhammes, A., and Jia, Y. (2007) Synthesis of graphene-based nanosheets via chemical reduction of exfoliated graphite oxide. *Carbon*, **45**, 1558–1565.
52. Eda, G., Fanchini, G., and Chhowalla, M. (2008) Large-area ultrathin films of reduced graphene oxide as a transparent and flexible electronic material. *Nat. Nanotechnol.*, **3**, 270–274.
53. Singh, V., Joung, D., Zhai, L., Das, S., Khondaker, S.I., and Seal, S. (2011) Graphene based materials: past, present and future. *Prog. Mater. Sci.*, **56**, 1178–1271.
54. Kuila, T., Mishra, A.K., Khanra, P., Kim, N.H., and Lee, J.H. (2013) Recent advances in the efficient reduction of graphene oxide and its application as energy storage electrode materials. *Nanoscale*, **5**, 52–71.
55. Incze, P.N., Osvatha, Z., Kamarasb, K., and Biro, L.P. (2008) Anomalies in thickness measurements of graphene and few layer graphite crystals by tapping mode atomic force microscopy. *Carbon*, **46**, 1435–1442.
56. Kudin, K.N., Scuseria, G.E., and Yakobson, B.I. (2001) Oxygen-driven unzipping graphite materials. *Phys. Rev. B*, **64**, 235406.
57. Du, X., Skachko, I., Barker, A., and Andrei, E.Y. (2008) Approaching ballistic transport in suspended graphene. *Nat. Nanotechnol.*, **3**, 491–495.
58. Morozov, S.V., Novoselov, K.S., Katsnelson, M.I., Schedin, F., Elias, D.C., and Jaszczak, J.A. (2008) Giant intrinsic carrier mobilities in graphene and its bilayer. *Phys. Rev. Lett.*, **100**, 016602.
59. Kuila, T., Bose, S., Mishra, A.K., Khanra, P., Kim, N.H., and Lee, J.H. (2012) Chemical functionalization of graphene and its applications. *Prog. Mater Sci.*, **57**, 1061–1105.
60. Lonkar, S.P., Deshmukh, Y.S., and Abdala, A.A. (2015) Recent advances in chemical modifications of graphene. *Nano Res.*, **8**, 1039–1074.
61. Georgakilas, V., Otyepka, M., Bourlinos, A.B., Chandra, V., Kim, N., Kemp, K.C., Hobza, P., Zboril, R., and Kim, K.S. (2012) Functionalization of graphene: covalent and non-covalent approaches, derivatives and applications. *Chem. Rev.*, **112**, 6156–6214.
62. Xu, Y., Liu, Z., Zhang, X., Wang, Y., Tian, J., and Huang, Y. (2009) A graphene hybrid material covalently functionalized with porphyrin: synthesis and optical limiting property. *Adv. Mater.*, **21**, 1275–1279.
63. Kang, S.M., Park, S., Kim, D., Park, S.Y., Ruoff, R.S., and Lee, H. (2011) Simultaneous reduction and surface functionalization of graphene oxide by

mussel-inspired chemistry. *Adv. Funct. Mater.*, **21**, 108–112.
64. Yu, W., Xie, H., Wang, X., and Wang, X. (2011) Highly efficient method for preparing homogeneous and stable colloids containing graphene oxide. *Nanoscale Res. Lett.*, **6**, 47.
65. Kim, N.H., Kuila, T., and Lee, J.H. (2013) Simultaneous reduction, functionalization and stitching of graphene oxide with ethylene diamine for composites application. *J. Mater. Chem. A*, **1**, 1349–1358.
66. Li, W., Tang, X.Z., Zhang, H.B., Jiang, Z.G., Yu, Z.Z., Du, X.S., and Mai, Y.W. (2011) Simultaneous surface functionalization and reduction of graphene oxide with octadecylamine for electrically conductive polystyrene composites. *Carbon*, **49**, 4724–4730.
67. Kuila, T., Khanra, P., Mishra, A.K., Kim, N.H., and Lee, J.H. (2012) Functionalized-graphene/ethylene vinyl acetate co-polymer composites for improved mechanical and thermal properties. *Polym. Test.*, **31**, 282–289.
68. Jang, J., Pham, V.H., Rajagopalan, B., Hur, S.H., and Chung, J.S. (2014) Effects of the alkylamine functionalization of graphene oxide on the properties of polystyrene nanocomposites. *Nanoscale Res. Lett.*, **9**, 265.
69. Stankovich, S., Piner, R.D., Nguyen, S.B.T., and Ruoff, R.S. (2006) Synthesis and exfoliation of isocyanate-treated graphene oxide nanoplatelets. *Carbon*, **44**, 3342–3347.
70. Liu, Z., Robinson, J.T., Sun, X.M., and Dai, H.J. (2008) PEGylated nano graphene oxide for delivery of water-insoluble cancer drugs. *J. Am. Chem. Soc.*, **130**, 10876–10877.
71. Bao, H., Pan, Y., Ping, Y., Sahoo, N.G., Wu, T., Li, L., Li, J., and Gan, L.H. (2011) Chitosan-functionalized graphene oxide as a nanocarrier for drug and gene delivery. *Small*, **7**, 1569–1578.
72. Salavagione, H.J., Gomez, M.A., and Martınez, G. (2009) Polymeric modification of graphene through esterification of graphite oxide and poly (vinyl alcohol). *Macromolecules*, **42**, 6331–6334.
73. Salavagione, H.J. and Martinez, G. (2011) Importance of covalent linkages in the preparation of effective reduced graphene oxide-poly(vinyl chloride) nanocomposites. *Macromolecules*, **44**, 2685–2692.
74. Zhang, Y., Ren, L., Wang, S., Marathe, A., Chaudhuri, J., and Li, G. (2011) Functionalization of graphene sheets through fullerene attachment. *J. Mater. Chem.*, **21**, 5386–5391.
75. Wang, X., Xing, W., Zhang, P., Song, L., Yang, H., and Hu, Y. (2012) Covalent functionalization of graphene with organosilane and its use as a reinforcement in epoxy composites. *Compos. Sci. Technol.*, **72**, 737–743.
76. Fang, M., Wang, K., Lu, H., Yanga, Y., and Nutt, S. (2009) Covalent polymer functionalization of graphene nanosheets and mechanical properties of composites. *J. Mater. Chem.*, **19**, 7098–7105.
77. Shen, J., Hu, Y., Li, C., Qin, C., and Ye, M. (2009) Synthesis of amphiphilic graphene nanoplatelets. *Small*, **5**, 82–85.
78. Lee, S.H., Dreyer, D.R., An, J., Velamakanni, A., Piner, R.D., and Park, S. (2010) Polymer brushes via controlled, surface initiated atom transfer radical polymerization (ATRP) from graphene oxide. *Macromol. Rapid Commun.*, **31**, 281–288.
79. Layek, R.K., Samanta, S., Chatterjee, D.P., and Nandi, A.K. (2010) Physical and mechanical properties of poly(methylmethacrylate)-functionalized graphene/poly(vinylidinefluoride) nanocomposites: piezoelectric polymorph formation. *Polymer*, **51**, 5846–5856.
80. Xu, Y.X., Bai, H., Lu, G.W., Li, C., and Shi, G.Q. (2008) Flexible graphene films via the filtration of water-soluble non-covalent functionalized graphene sheets. *J. Am. Chem. Soc.*, **130**, 5856–5857.
81. Liang, J., Huang, Y., Zhang, L., Wang, Y., Ma, Y., and Guo, T. (2009) Molecular-level dispersion of graphene into poly(vinyl alcohol) and effective

reinforcement of their nanocomposites. *Adv. Funct. Mater.*, **19**, 2297–2302.
82. Jana, M., Saha, S., Khanra, P., Samanta, P., Kood, H.Y., Murmu, N.C., and Kuila, T. (2015) Non-covalent functionalization of reduced graphene oxide using sulfanilic acid azocromotrop and its application as a supercapacitor electrode material. *J. Mater. Chem. A*, **3**, 7323–7331.
83. Teng, C.C., Ma, C.C.M., Lu, C.H., Yang, S.Y., Lee, S.H., Hsiao, M.C., Yen, M.Y., Chiou, K.C., and Lee, T.M. (2011) Thermal conductivity and structure of non-covalent functionalized graphene/epoxy composites. *Carbon*, **49**, 5107–5116.
84. Kuila, T., Mishra, A.K., Khanra, P., Kim, N.H., Uddin, M.E., and Lee, J.H. (2012) Facile method for the preparation of water dispersible graphene using sulfonated poly(ether-ether-ketone) and its application as energy storage materials. *Langmuir*, **28**, 9825–9833.
85. Layek, R.K., Das, A.K., Park, M.J., Kim, N.H., and Lee, J.H. (2015) Enhancement of physical, mechanical, and gas barrier properties in noncovalently functionalized graphene oxide/poly(vinylidene fluoride) composites. *Carbon*, **81**, 329–338.
86. Sun, X., Sun, H., Li, H., and H, P. (2013) Developing polymer composite materials: carbon nanotubes or graphene? *Adv. Mater.*, **25**, 5153–5176.
87. Zheng, W., Lu, X., and Wong, S.C. (2004) Electrical and mechanical properties of expanded graphite-reinforced high-density polyethylene. *J. Appl. Polym. Sci.*, **91**, 2781–2788.
88. Ye, L., Meng, X.Y., Ji, X., Li, Z.M., and Tang, J.H. (2009) Synthesis and characterization of expandable graphite-poly(methyl methacrylate) composite particles and their application to flame retardation of rigid polyurethane foams. *Polym. Degrad. Stab.*, **94**, 971–979.
89. Chen, G., Wu, D., Weng, W., and Wu, C. (2003) Exfoliation of graphite flakes and its nanocomposites. *Carbon*, **41**, 619–621.
90. Hu, H.T., Wang, J.C., Wan, L., Liu, F.M., and Zheng, H. (2010) Preparation and properties of graphene nanosheets—polystyrene nanocomposites via in situ emulsion polymerization. *Chem. Phys. Lett.*, **484**, 247–253.
91. Lianga, J., Wanga, Y., Huanga, Y., Maa, Y., Liua, Z., and Caib, J. (2009) Electromagnetic interference shielding of graphene/epoxy composites. *Carbon*, **47**, 922–925.
92. Kornmann, X. (2001) Synthesis and characterization of thermoset-layered silicate nanocomposites. PhD thesis. Sweden: Lulea Tekniska Universitet.
93. Moujahid, E.M., Besse, J.P., and Leroux, F. (2003) Poly (styrene sulfonate) layered double hydroxide nanocomposites stability and subsequent structural transformation with changes in temperature. *J. Mater. Chem.*, **13**, 258–264.
94. Potts, J.R., Lee, S.H., Alam, T.M., An, J., Stoller, M.D., Piner, R.D., and Ruoff, R.S. (2011) Thermomechanical properties of chemically modified graphene/poly(methyl methacrylate) composites made by in situ polymerization. *Carbon*, **49**, 2615–2623.
95. Liang, J., Xu, Y., Huang, Y., Zhang, L., Wang, Y., Ma, Y., Li, F., Guo, T., and Chen, Y. (2009) Infrared-triggered actuators from graphene-based nanocomposites. *J. Phys. Chem.*, **113**, 9921–9927.
96. J. Du, H. M. Cheng, The fabrication, properties, and uses of graphene/polymer composites. *Macromol. Chem. Phys.* 2012, **213**, 1060–1077.
97. Kuilla, T., Srivastava, S.K., and Bhowmick, A.K. (2009) Rubber/LDH nanocomposites by solution blending. *J. Appl. Polym. Sci.*, **111**, 635–641.
98. Goncalves, G., Marques, P.A.A.P., Timmons, A.B., Bdkin, I., Singh, M.K., Emamic, N., and Gracio, J. (2010) Graphene oxide modified with PMMA via ATRP as reinforcement filler. *J. Mater. Chem.*, **20**, 9927–9934.
99. Lee, Y.R., Raghu, A.V., Jeong, H.M., and Kim, B.K. (2009) Properties of waterborne polyurethane/functionalized

graphene sheet nanocomposites prepared by an in situ method. *Macromol. Chem. Phys.*, **210**, 1247–1254.
100. Kuila, T., Bose, S., Hong, C.E., Uddin, M.E., Khanra, P., Kim, N.H., and Lee, J.H. (2011) Preparation of functionalized graphene/linear low density polyethylene composites by a solution mixing method. *Carbon*, **49**, 1033–1051.
101. Kalaitzidou, K., Fukushima, H., and Drzal, L.T. (2007) A new compounding method for exfoliated graphite-polypropylene nanocomposites with enhanced flexural properties and lower percolation threshold. *Compos. Sci. Technol.*, **67**, 2045–2051.
102. Kim, I.H. and Jeong, Y.G. (2010) Polylactide/exfoliated graphite nanocomposites with enhanced thermal stability, mechanical modulus, and electrical conductivity. *J Polym. Sci. B Polym. Phys.*, **48**, 850–858.
103. Zhang, H.B., Zheng, W.G., Yan, Q., Yang, Y., Wang, J.W., Lu, Z.H., Yu, G.Y., and Ji, Z.Z. (2010) Electrically conductive polyethylene terephthalate/graphene nanocomposites prepared by melt compounding. *Polymer*, **51**, 1191–1196.
104. Song, P., Cao, Z., Cai, Y., Zhao, L., Fang, Z., and Fu, S. (2011) Fabrication of exfoliated graphene-based polypropylene nanocomposites with enhanced mechanical and thermal properties. *Polymer*, **52**, 4001–4010.
105. Lin, Y., Jin, J., and Song, M. (2011) Preparation and characterisation of covalent polymer functionalized graphene oxide. *J. Mater. Chem.*, **21**, 3455–3461.
106. Yang, M., Hou, Y., and Kotov, N.A. (2012) Graphene-based multilayers: critical evaluation of materials assembly techniques. *Nano Today*, **7**, 430–447.
107. Zhao, X., Zhang, Q.H., Hao, Y.P., Li, Y.Z., Fang, Y., and Chen, D.J. (2010) Enhanced mechanical properties of graphene based poly(vinyl alcohol) composites. *Macromolecules*, **43**, 9411–9416.
108. Kulkarni, D.D., Choi, I., Singamaneni, S.S., and Tsukruk, V.V. (2010) Graphene oxide-polyelectrolyte nano membranes. *ACS Nano*, **4**, 4667–4676.
109. Layek, R.K., Samanta, S., and Nandi, A.K. (2012) The physical properties of sulfonated graphene/poly(vinyl alcohol) composites. *Carbon*, **50**, 815–827.
110. Saini, P., Kuila, T., Saha, S., and Murmu, N.C. (2014) Graphene and its nanocomposites for gas sensing applications, in *Advanced Carbon Materials and Technology*, Scrivener Publishing LLC, pp. 501–534.
111. Kuila, T., Bose, S., Mishra, A.K., Khanra, P., Kim, N.H., and Lee, J.H. (2012) Effect of functionalized graphene on the physical properties of linear low density polyethylene nanocomposites. *Polym. Test.*, **31**, 31–38.
112. Rafiee, M.A., Lu, W., Thomas, A.V., Zandiatashbar, A., Rafiee, J., Tour, J.M., and Koratkar, N.A. (2010) Graphene nanoribbon composites. *ACS Nano*, **4**, 7415–7420.
113. Wang, J., Wang, X., Xu, C., Zhang, M., and Shang, X. (2011) Preparation of graphene/poly(vinyl alcohol) nanocomposites with enhanced mechanical properties and water resistance. *Polym. Int.*, **60**, 816–822.
114. Cano, M., Khan, U., Sainsbury, T., Neill, A.O., Wang, Z., McGovern, I.T., Maser, W.K., Benito, A.M., and Coleman, J.N. (2013) Improving the mechanical properties of graphene oxide based materials by covalent attachment of polymer chains. *Carbon*, **52**, 363–371.
115. Vadukumpully, S., Paul, J., Mahanta, N., and Valiyaveettil, S. (2011) Flexible conductive graphene/poly(vinyl chloride) composite thin films with high mechanical strength and thermal stability. *Carbon*, **49**, 198–205.
116. Zhao, X., Zhang, Q., and Chen, D. (2010) Enhanced mechanical properties of graphene-based poly (vinyl alcohol) composites. *Macromolecules*, **43**, 2357–2363.
117. Hwang, S., Kang, D., Ruoff, R.S., Shin, H.S., and Park, Y. (2014) Poly(vinyl alcohol) reinforced and toughened with poly(dopamine)-treated graphene oxide, and its use for humidity sensing. *ACS Nano*, **8**, 6739–6747.

118. Istrate, O.M., Paton, K.R., Khan, U., O'Neill, A., Bell, A.P., and Coleman, J.N. (2014) Reinforcement in melt-processed polymer-graphene composites at extremely low graphene loading level. *Carbon*, **78**, 243–249.
119. Shim, S.H., Kim, K.T., Lee, J.U., and Jo, W.H. (2012) Facile method to functionalize graphene oxide and its application to poly(ethylene terephthalate)/graphene composite. *ACS Appl. Mater. Interfaces*, **4**, 4184–4191.
120. Liao, W.H., Yang, S.Y., Hsiao, S.T., Wang, Y.S., Li, S.M., Ma, C.C.M., Tien, H.W., and Zeng, S.J. (2014) Effect of octa(aminophenyl) polyhedral oligomeric silsesquioxane functionalized graphene oxide on the mechanical and dielectric properties of polyimide composites. *ACS Appl. Mater. Interfaces*, **6**, 15802–15812.
121. Wang, J., Yang, S., Huang, Y., Tien, H., Chin, W., and Ma, C.M. (2011) Preparation and properties of graphene oxide/polyimide composite films with low dielectric constant and ultrahigh strength via in situ polymerization. *J. Mater. Chem.*, **21**, 13569–13575.
122. Liao, W., Yang, S., Wang, J., Tien, H., Hsiao, S., Wang, Y., Li, S., Ma, C.M., and Wu, Y. (2013) Effect of molecular chain length on the mechanical and thermal properties of amine-functionalized graphene oxide/polyimide composite films prepared by in situ polymerization. *ACS Appl. Mater. Interfaces*, **5**, 869–877.
123. Cai, D., Jin, J., Yusoh, K., Rafiq, R., and Song, M. (2012) High performance polyurethane/functionalized graphene nanocomposites with improved mechanical and thermal properties. *Compos. Sci. Technol.*, **72**, 702–707.
124. Xiang, C., Cox, P.J., Kukovecz, A., Genorio, B., Yan, D.P.H.Z., Peng, Z., Hwang, C., Ruan, G., Samuel, E.L.G., Sudeep, P.M., Konya, Z., Vajtai, X., Ajayan, P.M., and Tour, J.M. (2013) Functionalized low defect graphene nanoribbons and polyurethane composite film for improved gas barrier and mechanical performances. *ACS Nano*, **7**, 10380–10386.
125. Wang, X., Hu, Y., Song, L., Yang, H., Xing, W., and Lu, H. (2011) In situ polymerization of graphene nanosheets and polyurethane with enhanced mechanical and thermal properties. *J. Mater. Chem.*, **21**, 4222–4227.
126. Jin, J., Rafiq, R., Gill, Y.Q., and Song, M. (2013) Preparation and characterization of high performance of graphene/nylon nanocomposites. *Eur. Polym. J.*, **49**, 2617–2626.
127. Roy, S., Tang, X., Das, T., Zhang, L., Li, Y., Ting, S., Hu, X., and Yue, C.Y. (2015) Enhanced molecular level dispersion and interface bonding at low loading of modified graphene oxide to fabricate super nylon 12 composites. *ACS Appl. Mater. Interfaces*, 7, 3142–3151.
128. Rafiee, M.A., Rafiee, J., Wang, Z., Song, H., Yu, Z.Z., and Koratkar, N. (2009) Enhanced mechanical properties of nanocomposites at low graphene content. *ACS Nano*, **3**, 3884–3890.
129. Wajid, A.S., Tanvir, A.H.S., Das, S., Irin, F., Jankowski, A.F., and Green, M. (2013) High-performance pristine graphene/epoxy composites with enhanced mechanical and electrical properties. *Macromol. Mater. Eng.*, **298**, 339–347.
130. Ye, S., Feng, J., and Wu, P. (2013) Deposition of three-dimensional graphene aerogel on nickel foam as a binder-free supercapacitor electrode. *ACS Appl. Mater. Interfaces*, **5**, 7122–7129.
131. Tang, B., Hu, G., Gao, H., and Shi, Z. (2013) Three-dimensional graphene network assisted high performance dye sensitized solar cells. *J. Power Sources*, **234**, 60–68.
132. Sun, H., Xu, Z., and Gao, C. (2013) Multifunctional, ultra-flyweight, synergistically assembled carbon aerogels. *Adv. Mater.*, **25**, 2554–2560.
133. Ni, Y., Chen, L., Teng, K., Shi, J., Qian, X., Xu, Z., Tian, X., Hu, C., and Ma, M. (2015) Superior mechanical properties of epoxy composites reinforced by 3D interconnected graphene skeleton. *ACS Appl. Mater. Interfaces*, 7, 11583–11591.

134. Jiang, T., Kuila, T., Kim, N.H., and Lee, J.H. (2014) Effects of surface-modified silica nanoparticles attached graphene oxide using isocyanate terminated flexible polymer chains on the mechanical properties of epoxy composites. *J. Mater. Chem. A*, **2**, 10557–10567.

135. Jiang, T., Kuila, T., Kim, N.H., Ku, B.C., and Lee, J.H. (2013) Enhanced mechanical properties of silanized silica nanoparticle attached graphene oxide/epoxy composites. *Compos. Sci. Technol.*, **79**, 115–125.

136. Tang, L.C., Wan, Y.J., Yan, D., Pei, Y.B., Zhao, L., Li, Y.B., Wu, L.B., Jiang, J.X., and Lai, G.Q. (2013) The effect of graphene dispersion on the mechanical properties of graphene/epoxy composites. *Carbon*, **60**, 16–27.

137. Jia, J., Sun, X., Lin, X., Shen, X., Mai, Y.W., and Kim, J.K. (2014) Exceptional electrical conductivity and fracture resistance of 3D interconnected graphene foam/epoxy composites. *ACS Nano*, **8**, 5774–5783.

138. Gao, J., Chen, F., Wang, K., Deng, H., Zhang, Q., Bai, H., and Fu, Q. (2011) A promising alternative to conventional polyethylene with poly (propylene carbonate) reinforced by graphene oxide nanosheets. *J. Mater. Chem.*, **21**, 17627–17630.

139. Layek, R.K., Kundu, A., and Nandi, A.K. (2013) High-performance nanocomposites of sodium carboxymethyl cellulose and graphene oxide. *Macromol. Mater. Eng.*, **298**, 1166–1175.

140. Zhang, Q., Xu, X., Li, H., Xiong, G., Fisher, T.S., and Hu, H. (2015) Mechanically robust honeycomb graphene aerogel multifunctional polymer composites. *Carbon*, **93**, 659–670.

141. Li, Q., Guo, Y., Li, W., Qiu, S., Zhu, C., Wei, X., Chen, M., Liu, C., Liao, S., Gong, Y., Mishra, A.K., and Liu, L. (2014) Ultrahigh thermal conductivity of assembled aligned multilayer graphene/epoxy composite. *Chem. Mater.*, **26**, 4459–4465.

142. Shtein, M., Nadiv, R., Buzaglo, M., Kahil, K., and Regev, O. (2015) Thermally conductive graphene-polymer composites: size, percolation, and synergy effects. *Chem. Mater.*, **27**, 2100–2106.

143. Song, S.H., Park, K.H., Kim, B.H., Choi, Y.W., Jun, G.H., Lee, D.J., Kong, B.S., Paik, K.W., and Jeon, S. (2013) Enhanced thermal conductivity of epoxy-graphene composites by using non-oxidized graphene flakes with non-covalent functionalization. *Adv. Mater.*, **25**, 732–737.

144. Wang, S., Tambraparni, M., Qiu, J., Tipton, J., and Dean, D. (2009) Thermal expansion of graphene composites. *Macromolecules*, **42**, 5251–5255.

145. Fang, M., Wang, K., Lu, H., Yanga, Y., and Nutt, S. (2010) Single-layer graphene nanosheets with controlled grafting of polymer chains. *J. Mater. Chem.*, **20**, 1982–1992.

146. Araby, S., Zhang, L., Kuan, H., Dai, J., Majewski, P., and Maa, J. (2013) A novel approach to electrically and thermally conductive elastomers using graphene. *Polymer*, **54**, 3663–3670.

147. Patole, A.S., Patole, S.P., Kang, H., Yoo, J., Kim, T., and Ahn, J. (2010) A facile approach to the fabrication of graphene/polystyrene nanocomposites by in situ micro emulsion polymerization. *J. Colloid Interface Sci.*, **350**, 530–537.

148. Salavagione, H.J., Martınez, G., and Gomez, M.A. (2009) Synthesis of poly (vinyl alcohol)/reduced graphite oxide nanocomposites with improved thermal and electrical properties. *J. Mater. Chem.*, **19**, 5027–5032.

149. Kuila, T., Bose, S., Khanra, P., Kim, N.H., Rhee, K.Y., and Lee, J.H. (2011) Characterization and properties of in situ emulsion polymerized poly(methyl methacrylate)/graphene nanocomposites. *Compos. Part A*, **42**, 1856–1861.

150. Yun, Y.S., Bae, Y.H., Kim, D.H., Lee, J.Y., Chin, I., and Jin, H. (2011) Reinforcing effects of adding alkylated graphene oxide to polypropylene. *Carbon*, **49**, 3553–3559.

151. Stankovich, S., Dikin, D.A., Dommett, G.H.B., Kohlhaas, K.M., Zimney, E.J., Stach, E.A., Pine, R.D., Nguyen, S.B.T., and Ruoff, R.S. (2006) Graphene-based composite materials. *Nature*, **442**, 282–286.

152. Pham, V.H., Cuong, T.V., Dang, T.T., Hur, S.H., Kim, B.K.E.J., Shin, E.W., and Chung, J.S. (2011) Superior conductive polystyrene-chemically converted graphene nanocomposite. *J. Mater. Chem.*, **21**, 11312–11316.
153. Qi, X.Y., Yan, D., Jiang, Z., Cao, Y.K., Yu, Z.Z., Yavari, F., and Koratkar, N. (2011) Enhanced electrical conductivity in polystyrene composites at ultra-low graphene content. *ACS Appl. Mater. Interfaces*, **3**, 3130–3133.
154. Zheng, D., Tang, G., Zhang, H., Yu, Z., Yavari, F., Koratkar, N., Lim, S., and Lee, M. (2012) In situ thermal reduction of graphene oxide for high electrical conductivity and low percolation threshold in polyamide 6 nanocomposites. *Compos. Sci. Technol.*, **72**, 284–289.
155. Wu, C., Huang, X., Wang, G., Lv, L., Chen, G., Li, G., and Jiang, P. (2013) Highly conductive nanocomposites with three-dimensional, compactly interconnected graphene networks via a self-assembly process. *Adv. Funct. Mater.*, **23**, 506–513.
156. Hsiao, S., Ma, C.M., Liao, W., Wang, Y., Li, S., Huang, Y., Yang, R., and Liang, W. (2014) Lightweight and flexible reduced graphene oxide/water-borne polyurethane composites with high electrical conductivity and excellent electromagnetic interference shielding performance. *ACS Appl. Mater. Interfaces*, **6**, 10667–10678.
157. Zhang, H.B., Zheng, W.G., Yan, Q., Yang, Y., Wang, J.W., Lu, Z.H., Ji, G.Y., and Yu, Z.Z. (2011) Electrically conductive polyethylene terephthalate/graphene nanocomposites prepared by melt compounding. *Polymer*, **51**, 1191–1196.
158. Shen, Y., Jing, T., Ren, W., Zhang, J., Jiang, Z.G., Yu, Z.Z., and Dasari, A. (2012) Chemical and thermal reduction of graphene oxide and its electrically conductive polylactic acid nanocomposites. *Comp. Sci. Technol.*, **72**, 1430–1435.
159. Vuluga, D., Thomassin, J.M., Molenberg, I., Huynen, I., Gilbert, B., Jerome, C., Alexandrea, M., and Detrembleur, C. (2011) Straightforward synthesis of conductive graphene/polymer nanocomposites from graphite oxide. *Chem. Commun.*, **47**, 2544–2546.
160. Chen, Z., Ren, W., Gao, L., Liu, B., Pei, S., and Cheng, H.M. (2011) Three-dimensional flexible and conductive interconnected graphene networks grown by chemical vapour deposition. *Nat. Mater.*, **10**, 424–428.
161. Yoonessi, M. and Gaier, J.R. (2010) Highly conductive multifunctional graphene polycarbonate nanocomposites. *ACS Nano*, **4**, 7211–7220.
162. Vasileiou, A.A., Kontopoulou, M., and Docoslis, A. (2014) A Noncovalent Compatibilization Approach to Improve the Filler Dispersion and Properties of Polyethylene/Graphene Composites. *ACS Appl. Mater. Interfaces*, **6**, 1916–1925.
163. Qi, X.Y., Yan, D., Jiang, Z., Cao, Y.K., Yu, Z.Z., Yavari, F., and Koratkar, N. (2011) Enhanced electrical conductivity in polystyrene composites at ultra-low graphene content. *ACS Appl. Mater. Interfaces*, **3**, 3130–3133.
164. Hyunwoo, H., Zhang, G., Xiao, L., Wang, H., Zhang, Q., and Zhao, Z. (2012) Preparation and electrical conductivity of graphene/ultrahigh molecular weight polyethylene composites with a segregated structure. *Carbon*, **50**, 4596–4599.
165. Kim, H., Miura, Y., and Macosko, C.W. (2010) Graphene/polyurethane nanocomposites for improved gas barrier and electrical conductivity. *Chem. Mater.*, **22**, 3441–3450.
166. Mittal, V. and Chaudhry, A.U. (2015) Polymer–graphene nanocomposites: effect of polymer matrix and filler amount on properties. *Macromol. Mater. Eng.*, **300**, 510–521.
167. Yoonessi, M., Shi, Y., Scheiman, D.A., Colon, M.L., Tigelaar, D.M., Weiss, R.A., and Meador, M.A. (2012) Graphene polyimide nanocomposites; thermal, mechanical, an high-temperature shape memory effects. *ACS Nano*, **6**, 7644–7655.
168. Yang, Y., Bolling, L., Priolo, M.A., and Grunlan, J.C. (2013) Super gas barrier

and selectivity of graphene oxide-polymer multilayer thin films. *Adv. Mater.*, **25**, 503–508.

169. Liu, H., Kuila, T., Kim, N.H., Kud, B., and Lee, J.H. (2013) In situ synthesis of the reduced graphene oxide–polyethyleneimine composite and its gas barrier properties. *J. Mater. Chem. A*, **1**, 3739–3746.

170. Wu, Z., Chen, X., Zhu, S., Zhou, Z., Yao, Y., Quan, W., and Liu, B. (2013) Enhanced sensitivity of ammonia sensor using graphene/polyaniline nanocomposite. *Sens. Actuators B: Chem.*, **178**, 485–493.

171. Yang, G., Lee, C., Kim, J., Ren, F., and Pearton, S.J. (2013) Flexible graphene-based chemical sensors on paper substrates. *Phys. Chem. Chem. Phys.*, **15**, 1798.

172. Bai, H., Sheng, K., Zhang, P., Li, C., and Shi, G. (2011) Graphene oxide/conducting polymer composite hydrogels. *J. Mater. Chem.*, **21**, 18653–18658.

173. Yan, D., Pang, H., Li, B., Vajtai, R., Xu, L., Ren, P., Wang, J., and Li, Z. (2015) Structured reduced graphene oxide/polymer composites for ultra-efficient electromagnetic interference shielding. *Adv. Funct. Mater.*, **25**, 559–566.

174. Eswaraiah, V., Sankaranarayanan, V., and Ramaprabh, S. (2011) Functionalized graphene/PVDF foam composites for EMI shielding. *Macromol. Mater. Eng.*, **296**, 894–898.

175. Wang, X., Song, L., Yang, H., Xing, W., Kandol, B., and Hu, Y. (2012) Simultaneous reduction and surface functionalization of graphene oxide with POSS for reducing fire hazards in epoxy composites. *J. Mater. Chem.*, **22**, 22037.

176. Bao, C., Guo, Y., Yuan, B., and Lei Song, Y.H. (2012) Functionalized graphene oxide for fire safety applications of polymers: a combination of condensed phase flame retardant strategies. *J. Mater. Chem.*, **22**, 23057.

177. V. Patel, Y. Mahajan, Polymer nanocomposites drive opportunities in the automotive sector.

178. http://www.graphene-info.com/composite-materials-study-can-boos (accessed 29 February 2016).

179. Song L., Guo, J., and Zhamu, A. (2009) Highly conductive nano scaled graphene plate nanocomposites. US Patent 7566410.

180. Grujicic, M., Arakere, G., and Pandurangan, B. (2010) Multidisciplinary design optimization for glass-fiber epoxy-matrix composite 5MW horizontal-axis wind turbine blades. *J. Mater. Eng. Perform.*, **19**, 1116–1127.

181. Al-Qabandi, O., Silva, A.D., Al-Enezi, S., and Bassyouni, M. (2014) Synthesis, fabrication and mechanical characterization of reinforced epoxy and polypropylene composites for wind turbine blades. *J. Reinf. Plast. Compos.*, **33**, 2287–2299.

4
Graphene in Lithium-ion Batteries

Cyrus Zamani

4.1
Introduction

Energy is one of the main concerns of our era. There are several reasons for this, among them, the increased population of the planet and, therefore, the need for more energy, the high level of pollution, and global warming. On one hand, the worldwide energy demand is growing continuously and, on the other hand, environmental issues related to the current energy resources and technologies raise a big concern about the future of human beings. To cope with the existing problems and design a better future for the inhabitants of this planet, researchers are working actively in close collaboration with governments and policy makers. Figures and statistics show that still about 80% of the global energy demand is covered by fossil fuels and the same trend is expected until 2040 (Figure 4.1) although the demand will rise to one-third (from 500 EJ in 2008 to about 750 EJ by 2040) [1]. However, the conventional fuels are expected to run out in the near future (Table 4.1) [2] and, therefore, there is a great interest on using other sources of energy to answer the growing energy demand although the cumulative $7.4 trillion by 2040 looks to be still insufficient.

Looking at the global investment in the field of renewable energies, its importance is disclosed. According to UNEP [3], in 2014, $270.2 billion was invested on renewable energies and fuels, about 17% higher than 2013. Developing countries such as Chile, Mexico, Kenya, South Africa, and Turkey are among those who spent in the order of billion dollars. Jordan, Uruguay, Panama, The Philippines, and Myanmar also invested on renewable energies (in the range of $500 million to $1 billion). In total, about 48% of GW capacities installed in 2014 were renewables increasing the share of electricity from renewable sources to 9.1%. Of these, many installations are on-site and generate off-grid electricity, a fact that highlights the need to have storage devices for the produced electricity. High-performance batteries are therefore required for such systems distributed worldwide. Advanced lithium-ion batteries are considered to be the best candidates for such applications. Thanks to nanomaterials and nanotechnology,

Graphene Technology: From Laboratory to Fabrication,
First Edition. Edited by Soroush Nazarpour and Stephen R. Waite.

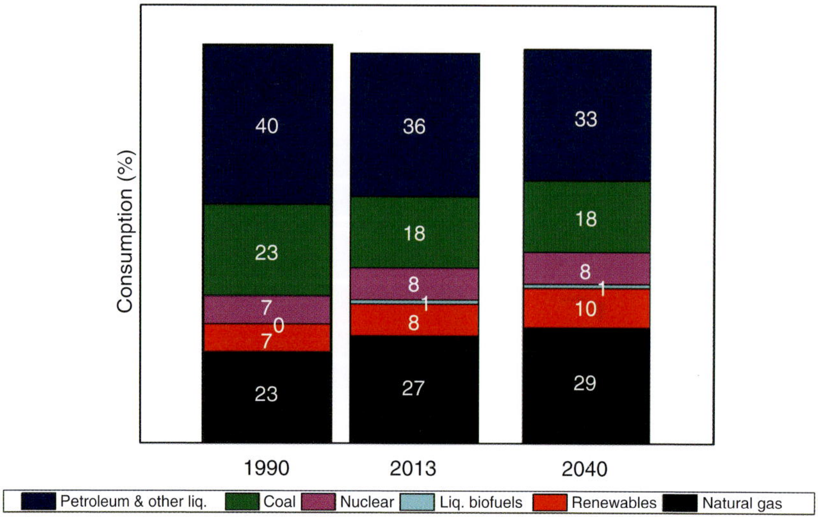

Figure 4.1 Evolution of energy consumption by resource type [1].

Table 4.1 The availability of fuels [2].

	Reserves (years)	Resources (years)
Oil (conventional)	43	67
Oil (conventional + Non-conventional)	62	157
Gas (conventional)	64	149
Gas (conventional + Non-conventional)	64	756
Hard coal	207	1425
Soft coal	198	1264
Uranium	42	527

these devices are gaining more and more importance in today's electronics. This chapter deals with the role of graphene materials in lithium-ion batteries.

4.2 Renewable Energies

Energy sources can be categorized in different ways. The most widely accepted one is, perhaps, according to their potentiality in being regenerated or not. In this aspect, energies are classified as non-renewables and renewables. The former mainly consists of fossil fuels (i.e., oil, gas, and coke) while the latter includes energy from sources such as sunlight, wind, biomass, and so on. Renewable energies are therefore named as clean or green energy as they do not have any negative impact on the environment and living things.

There exist numerous types of renewable energy sources. Basically, any source of energy that can be renewed in a very short period of time (in contrast to fossil sources that need thousands of years to form), falls in this category. Solar, tidal, biomass, wind, geothermal, and hydroenergy are common examples although there exist other sources that are less common (e.g., energy derived from algae, sugar, garbage, human fat, etc.,). With all their advantages, the global share of renewable energies is about 15% only, since these are not competitive in cost. No doubt, these emerging technologies come with their business risks, but they are accompanied with opportunities too. The fact is that if well established, the renewable energy industry will re-design the future of our planet and that is why the technology associated with renewable energies is considered as one of the great game players in tomorrow's technology sector. Whatever be the source of energy, there is always a need to store it for further use. This is more highlighted when we take into account the scattered nature of renewable energy production sites since, for example, wind turbines are in need of storing the produced energy from wind and this task is done by batteries implemented inside the machine or in the vicinity of the wind farm.

4.3
Batteries, What are They?

Batteries are devices used by mankind for energy storage. The energy is stored in the form of chemical energy, which is later converted to electricity to put a machine into operation. The oldest battery known is the Parthian battery (Figure 4.2) and dates back to the Sassanid period (about 250 BC) of the Persian Empire. Given this, batteries have a long history though the first mass-produced battery was released in 1802. Since then, different types of batteries (i.e., lead-acid, NiCd, lithium-ion, and NiMH) have been introduced (see Table 4.2 for a historical background on batteries and their evolution).

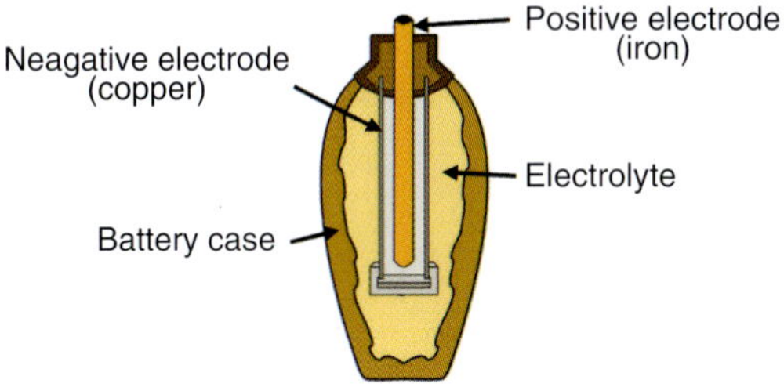

Figure 4.2 Parthian battery.

Table 4.2 Evolution of batteries.

Year/period	Battery type	Inventor
250 BC	Parthian battery	Sassanid Empire
1660	Electrical machine	Otto von Guericke
1744	Leyden jar	Ewald Georg von Kliest
1745–1827	Electric pistol	Alessandro Volta
1791	Galvanic electricity	Luigi Galvani
1800	Electrolysis	Sir HumphryDavi
1802	First electric battery (mass production)	William Cruickshank
1859	First rechargeable battery (lead acid)	Gaston Planté
1899	First nickel–cadmium battery	Waldmar Junger
1912	First Li-ion battery	G.N. Lewis
1947	Sealed NiCd cell	Georg Neumann
1970s	Non-rechargeable Li-ion battery	
1990s	Nickel-metal hydrides (NiMH)	
1991	Commercial rechargeable Li-ion battery	Sony Corporation

4.4 Lithium-ion Batteries

Lithium-ion batteries are a family of batteries that operate based on the intercalation of electrode structure with lithium ions. Today's batteries are rechargeable with lithium ions moving from the cathode (positive electrode) to the anode (negative electrode) during charging, as shown in Figure 4.3. The flow of ions is reversed during discharging (i.e., utilization) of the battery.

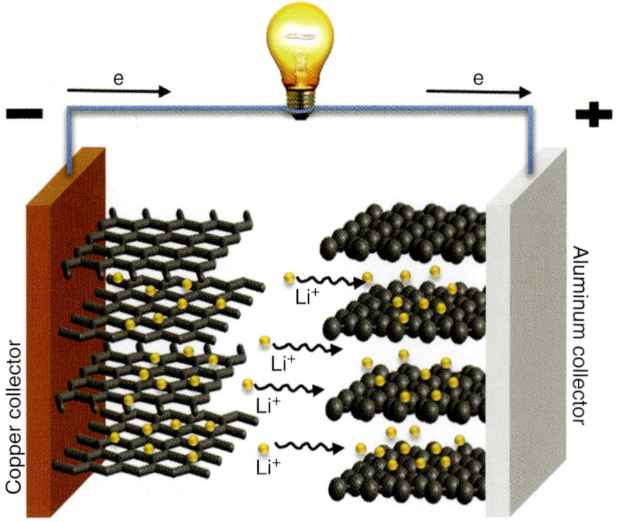

Figure 4.3 Principles of lithium-ion battery operation.

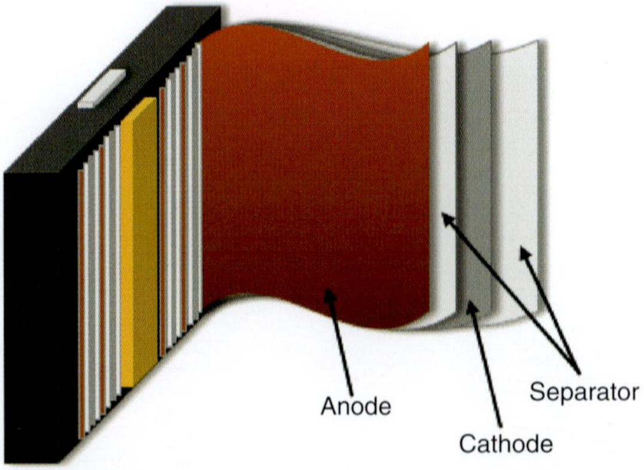

Figure 4.4 Schematic structure of a lithium-ion battery.

In comparison to other battery types, lithium-ion batteries offer higher energy density, longer cycle life, low self-discharge rate, and no memory effect. These advantages qualify lithium-ion batteries as the best choice for numerous applications including portable electronics. It is believed that lithium-ion batteries are environmentally friendly since no toxic metal is used in their structure and this makes them the best option for electric vehicles although some concern on the CO_2-emission of large battery packs is raised [4]. The main disadvantage is that these batteries are in need of special safety attention. Short-circuiting, overcharging, and overheating can arise which may lead to battery explosion. This makes a very careful design necessary.

All batteries have three parts in common, that is, anode, cathode, and electrolyte. Commercially available batteries are more sophisticated in structure. Basically, two common constructions are available: cylindrical and prismatic. Active materials (for cathode and anode) are deposited on aluminum and copper films. A four-layer spiral structure comprising the anode, cathode, and separators is put inside a package. In addition, there exist a variety of safety protection mechanisms including controller IC (to prevent overcharging), control switches (for charge/discharge interruption), temperature fuse (to prevent abnormal heating), and gas release control valve that is used to prevent battery explosion in case the internal gas pressure exceeds a critical value. The whole assembly is packed using a material (usually resin), which makes disassembling rather difficult (Figure 4.4).

4.5 Anodes, Cathodes, and Electrolytes

Commercially available lithium-ion batteries can be different in their electrode materials as well as electrolytes. Table 4.3 lists the common materials used as anode, cathode, and electrolyte. However, difficulties in fabrication of lithium-ion batteries with satisfactory cycling properties have made researchers to add

Table 4.3 Common materials used in lithium-ion batteries.

Anode	Graphite, hard carbon, LTO[a)]
Cathode	LCO, LFP, LMO, NCA, NMC[b)]
Electrolyte	Alkyl carbonates (EC, DMC, EMC, DEC), LiPF6[c)]

a) LTO: Li4Ti5O12.
b) LCO, $LiCoO_2$; LFP, $LiFePO_4$; LMO, $LiMn2O_4$; NCA, $LiNiCoAlO_2$; NMC, $LiNiMnCoO_2$.
c) EC, ethylene carbonate (EC); DMC, dimethyl carbonate; EMC, ethyl methyl carbonate; DEC, diethyl carbonate.

alloying compounds to the electrodes and create nanostructures for improved performance [5–7].

The electrolyte usually comprises lithium salt and transports lithium ions between the cathode and anode. An example of the electrolyte materials is LiPF6 in a solution to which other additives are also added for higher stability. The electrolytes are optimized for the specific anode and cathode materials used in battery structure.

For positive electrode, a number of materials are already in use. They usually have nickel, manganese, or cobalt in their crystal structure. Today, the main groups available in the market are lithium cobalt oxide (LCO), nickel manganese cobalt (NMC), lithium manganese oxide (LMO), and lithium-iron phosphate (LFP). The latter is gaining more importance due to its higher safety and low discharge rate.

In the case of negative electrode, so far, carbon has always been in use although recently, other materials such as silicon [8] and lithium titanate ($Li_4Ti_5O_{12}$) have also been announced [9].

For batteries with LCO (cathode), and graphite (anode), the chemical reactions would be:

$$LiCoO_2 \xrightarrow{\text{Charge}} Li_{1-x}CoO_2 + xLi^+ + xe^-$$

$$C + xLi^+ + xe^- \xrightarrow{\text{Charge}} CLi_x$$

$$LiCoO_2 + C \xrightarrow{\text{Charge}} Li_{1-x}CoO_2 + CLi_x$$

4.6 Carbon Materials

Graphite is the most common allotrope of carbon: an electrical conductor that has the highest stability among all carbon allotropes. The other well-known member of the family is diamond, a very hard material with light dispersion properties, which introduces the material as a good candidate for mechanical applications and in the manufacture of jewelry. In addition to graphite and diamond, there exist several other allotropes such as Lonsdaleite, nanotubes, fullerenes, graphene, and amorphous carbon.

4.7 Graphite

Graphite (Figure 4.5) – the most stable form of carbon – is known for its planar structure and relatively pure carbon composition. The material shows a combination of physical and chemical features. Graphite is electrically (and thermally) conductive. It is a corrosion-resistant material and so it is known as a chemically passive material (resistant to chemical attack). All these characteristics along with its low density, 2.266 g cm^{-3} (even lighter than aluminum) and high strength and stiffness make the material so special that in a wide variety of applications – from traditional furnace linings to advanced electrode materials for batteries – graphite is considered as one of the best candidates. In renewable energies and devices, fuel cells also benefit from graphite features in their bi-polar planes. In fact, the material is revolutionizing the world. The increasing consumer demand (both in traditional as well as advanced applications) forced global powers to classify graphite as a critical strategic material [10].

In lithium-ion batteries, both natural and synthetic graphite (obtained from mines and coke, respectively) are used. Natural graphite needs purification for being considered as an anode material for lithium-ion batteries. Currently, natural graphite is preferred for such applications since the material offers higher purity (99.9 vs. 99.0 in synthetic product) and higher degree of crystallinity (improved thermal and electrical conductivity) at a lower cost. As anode material, the morphology of graphite should be spherical (or potato type), Figure 4.6 [10].

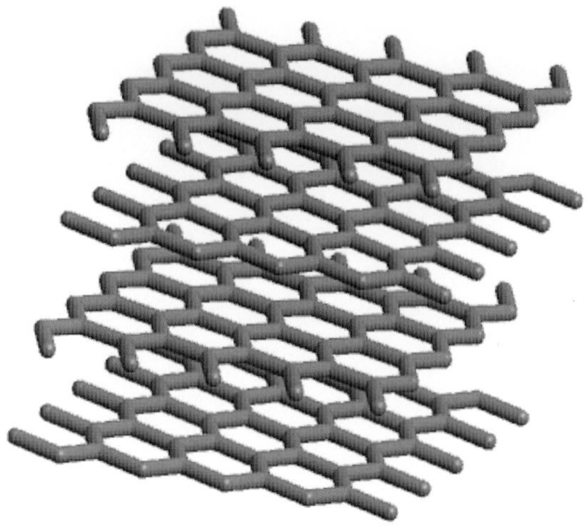

Figure 4.5 Structure of graphite.

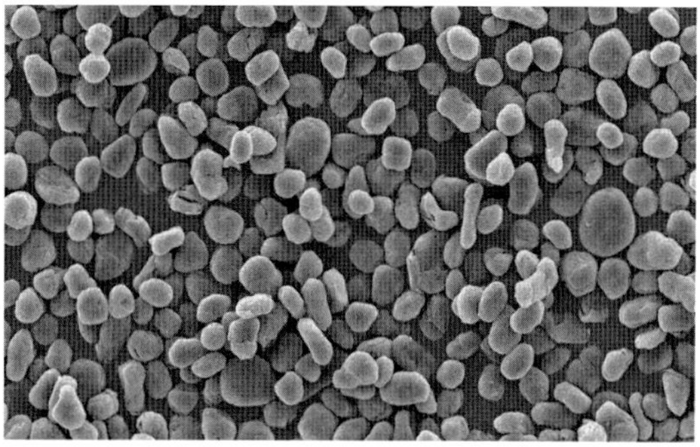

Figure 4.6 Battery grade graphite [11].

4.8 Graphene

Graphene is one of the greatest discoveries of materials scientists [12]. Discovered by A. Geim and K. Novoselov of the University of Manchester in 2004, it is a two-dimensional carbon allotrope with a hexagonal lattice (Figure 4.7), which can be prepared through a number of physical and/or chemical methods [13–15]. Graphene is the thinnest compound known so far comprising a single layer of carbon atoms that bond together (with sp2 hybridization) forming a tight honeycomb structure [16]. The bond length is 0.142 nm and the interlayer spacing (when stacked to form graphite) is 0.335 nm. Physically, graphene is the strongest material [17] with the highest room temperature and thermal and electrical conductivity (Table 4.4). Needless to say, these properties depend on the synthesis route and the structure of the material. For example, mechanically exfoliated products exhibit room temperature charge carrier mobility $>10^5\ cm^2\ V^{-1}\ s^{-1}$ while that of chemically exfoliated material (from graphene oxide) is about $1\ cm^2\ V^{-1}\ s^{-1}$ [18]. Moreover, the material is anisotropic with highest in-plane conductivity while in

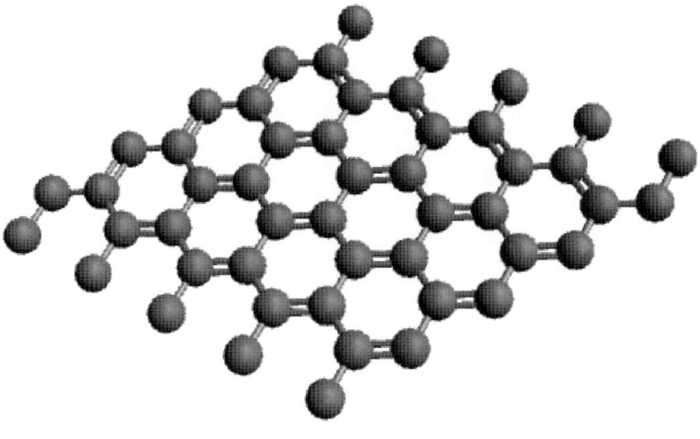

Figure 4.7 Graphene structure.

Table 4.4 Selected properties of graphene.

Property		Value
Dimensional	Thickness	3.35 A (varies depending on how layers are stacked)
	Specific surface area	$2630\ m^2\ g^{-1}$
Mechanical	Stiffness	150×10^6 PSi
	Young's modulus	1100 GPa
	Fracture strength	125 GPa
	Hardness	Highest among all carbon allotropes
	Tenacity	Flexible
Electronic		Zero-gap semiconductor, semi-metal
	Electron transport (RT)	$>15\,000\ cm^2\ V^{-1}\ s^{-1}$ (nearly identical electron & hole transport)
	Electrical conductivity	$2000\ S\ cm^{-1}$
Thermal	Thermal conductivity (RT)	$(4.84 \pm 0.44) \times 10^3$ to $(5.30 \pm 0.48) \times 10^3$ $W\ m^{-1}\ K^{-1}$
Optical	Transmittance	97.7%

the direction normal to the sheets, these properties drop to much lower values [19, 20]. On the other hand, with its basic hexagonal structure, graphene can act as the building block of other carbonaceous materials such as carbon nanotubes and fullerenes [21].

4.9 Graphene in Lithium-Ion Batteries

Graphene is a carbonaceous material with fantastic physical and chemical properties: highest electrical and thermal conductivity among all materials known to man, mechanical flexibility, thermal and chemical stability, extremely large surface area, and presence of proper sites for chemical adherence of materials. No other material with such a combination of properties can be found. All these features have attracted the interest of researchers for employing graphene in electrochemical devices such as lithium-ion batteries and fuel cells [13, 22, 23]. In lithium-ion batteries, for instance, graphene is a promising material since these properties make the insertion and extraction of Li^+ ions faster, easier, and reversible. In comparison to the rest of the members of the carbon family, graphene (in the form of graphene oxide and reduced graphene oxide (RGO)) has the advantage of possessing functional groups on the surface and edges. When making composites of graphene and metal oxides, for instance, the distribution of oxide particles as well as their size and shape are affected by these functional groups.

4.10 Graphene in Anodes

As a member of the carbon family, researchers started to explore the possibility of using graphene to improve the durability and charge capacity of lithium-ion battery anodes. In the case of graphite – which has been traditionally used as anode – the theoretical capacity is limited to 372 mA h g^{-1}, which is far below the needs of new devices (this capacity corresponds to intercalation of one lithium ion to six carbon atoms, LiC6) [24]. On an industrial scale, this deficiency is more pronounced since the microstructural features of carbon and graphite, and their crystallinity and morphology affect the functionality of the anode. The storage capacity of commercial devices is about 310 $mAhg^{-1}$ [25]. To go beyond this limit, different scenarios have been designed (Figure 4.8).

Modification of graphite by mild oxidation of the material induces micro porosities inside the material, thereby causing a better intercalation. On the other hand, a dense oxide layer is formed, which resists against electrode decomposition. The method, however, lacks reproducibility and uniformity in the final products, which is considered as a disadvantage [26].

The second strategy is based on creating composite structures using metals or metal oxides in a graphite matrix. Silver, nickel, and metal oxides such as SnO_2 are among the materials researched for this purpose. The idea behind this has been benefiting from the properties of the second phase (e.g., higher conductivity of silver and faster transport) to overcome the limitations of graphite. Making composite anodes is reported to enhance the electrode properties in several ways

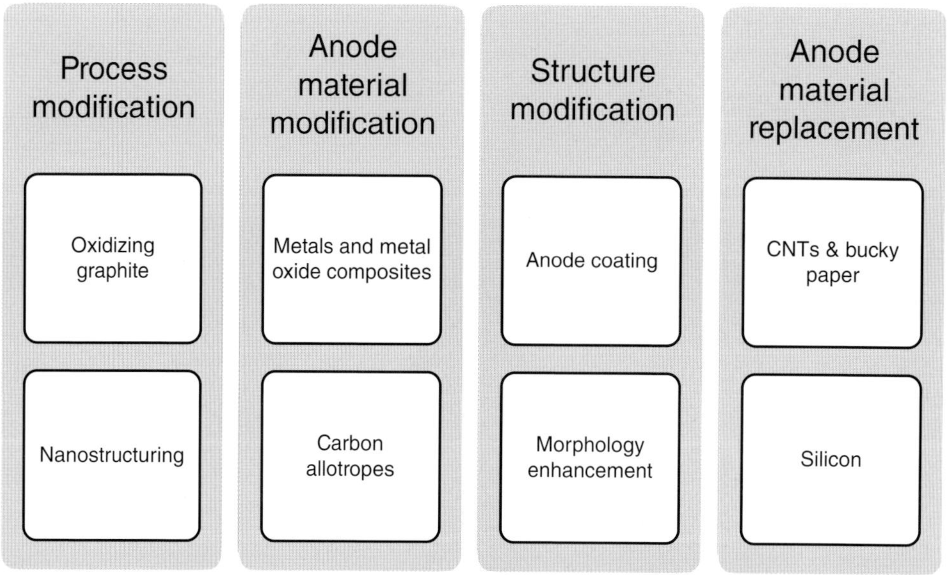

Figure 4.8 Different scenarios designed to improve anode performance.

including higher electrical conductivity, preventing the anode from decomposition, and creating hosts for lithium. The disadvantage, however, is the irreversible capacity noted in the first cycle [26]. Coating of anode with a polymer is another approach through which the anode material is coated with a polymer to improve its electrochemical performance. Both electronically conductive as well as ionically conductive polymers have been tried [27].

Despite all efforts made on the anode aspect of the lithium-ion batteries, still negative electrode is considered as a bottleneck restricting the device performance. Here, employing other nanostructured carbon allotropes would be a good strategy. While conventional carbons in the range of 0.1 – 10μm do not meet the demands, carbon nanotubes with diameters less than 100 nm could be a solution. The main advantage of carbon nanotubes is their ballistic electronic transport that ensures superior electrical conductivity [28, 29]. The major drawback of CNTs is limited intercalation since the only available sites are nanotube ends [30]. Moreover, nanostructuring can facilitate the charge/discharge processes since, according to Fick's second law of diffusion, a shorter diffusion distance results in shorter diffusion times [6].

While in graphite anodes, the intercalation is limited to the formation of LiC_6 [(0002) planes], graphene sheets expose both surfaces to lithium ions, thus providing the possibility of Li_2C_6 stoichiometry formation [24] and increasing the theoretical capacity to 740 mA h g^{-1}. Moreover, graphene offers additional sites (defects and imperfections) for ion storage, which therefore, ensure higher capacity for future batteries. Currently, various forms of graphene-based materials are examined for their usage in lithium-ion battery anodes. The major categories are listed in Figure 4.9 being graphene [22, 31], RGO [32], doped graphene [33, 34], and defected graphene sheets (DGNs) [35].

In comparison to graphene, RGO is always accompanied with a large amount of topological defects (mainly dislocations) and lack of crystallinity [36].

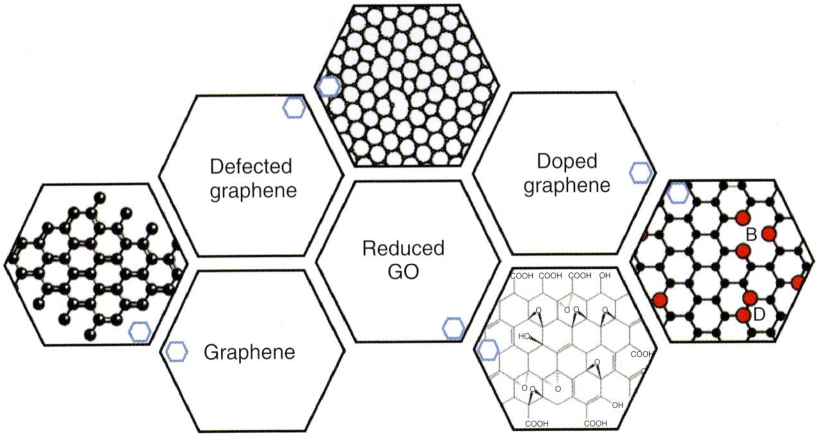

Figure 4.9 Graphene-based materials in lithium-ion battery anodes.

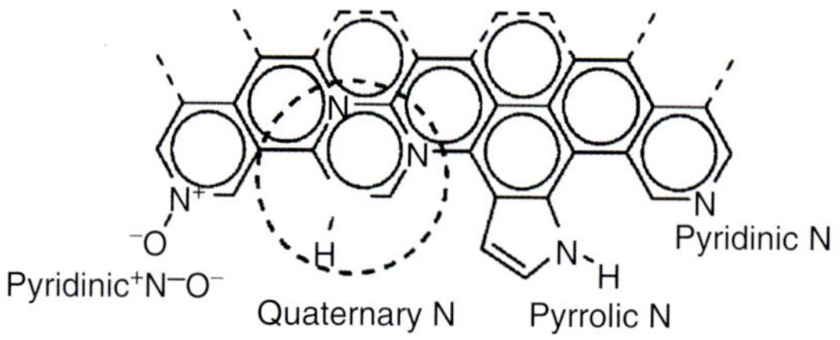

Figure 4.10 Bonding in nitrogen-doped graphene. (Reprinted from Ref. [33] with permission from the American Chemical Society, Copyright 2012.)

Therefore, it is expected to make use of such a structure in obtaining higher storage capacities. Heteroatom doping of graphene sheets with elements such as nitrogen, boron, phosphorus, and sulfur is also reported [21, 34]. Doping is done to improve graphene properties for devices such as lithium-ion batteries and supercapacitors. High-power and high-energy batteries featuring enhanced reversible charge/discharge are fabricated through doping. Wu *et al.* presented anode materials with reversible capacities above 1040 mA h g^{-1} [21]. Several nitrogen doping routes including CVD, thermal treatment, solvothermal, arc-discharge, plasma, and N2H4 treatment are reported as a result of which, three bonding configurations are obtained: pyridinic N, pyrrolic N, and graphitic N (Figure 4.10) [33]. Another way to create defects in graphene sheets is patterning their surface through processes such as lithography, electro plasma etching, and catalysis. Creating these defects provides the opportunity for the Li+ ions to penetrate into the structure, which is equivalent to higher capacities. Following a cobalt-catalyzed gasification strategy, Hu *et al.* obtained DGNs with a reversible specific capacity of about 1009 mA h g^{-1} [35].

All the above-mentioned graphene-based materials have shown great promise in enhancing lithium-ion battery operation. However, there remains a big challenge: the high tendency of graphene sheet to agglomeration and re-stacking due to van der Waals forces followed by a drop in electrical properties. To solve this serious issue, graphene composites are under research [37–45]. The behavior of graphene composites is a combination of the properties of both graphene and the other phase. In this respect, composites comprising graphene and metal oxides have shown great promise (Figure 4.11). The capacity of metal oxides is more than two times larger than that of graphite material and the reaction. For this, three mechanisms (conversion reaction, Li-alloy reaction, and Li insertion/extraction

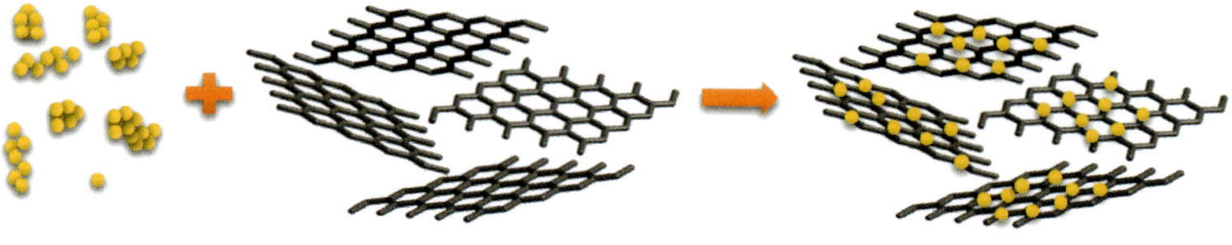

Figure 4.11 Graphene/MO composite formation. (Reproduced with permission from Ref. [21].)

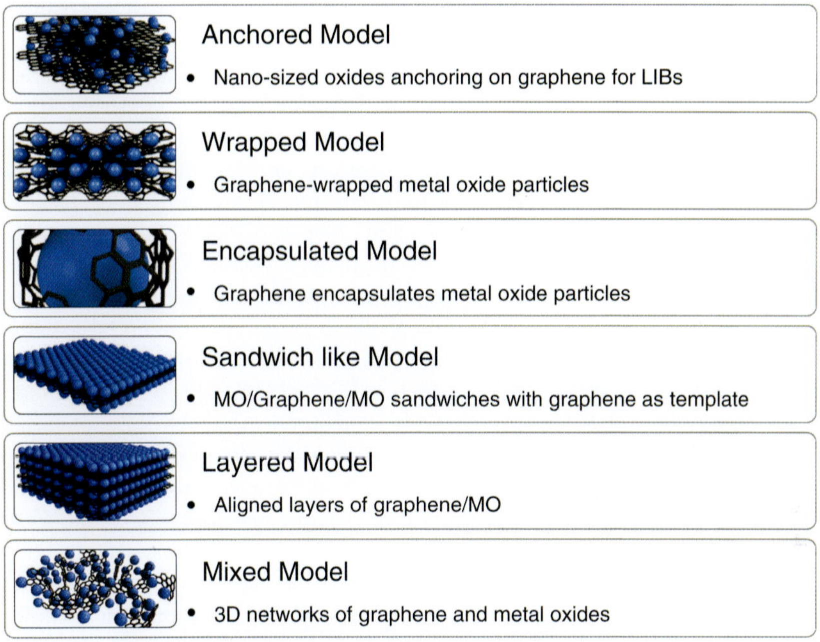

Figure 4.12 Structural mechanisms for compositing graphene with metal oxides for lithium battery applications. (Reproduced with permission from Ref. [23].)

reaction) are proposed and numerous metal oxides are tested. Examples are NiO, SnO_2, TiO_2, CuO, Cu_2O, MnO, Fe_2O_3, and Fe_3O_4 for which, six structural mechanisms are proposed (Figure 4.12). In all mechanisms, functional groups on the graphene surface and its edges result in improved chemical functionality and better compatibility to oxide particles. Based on their morphological features, on the other hand, metal oxide particles offer a high capacity. With these composite structures the following features are expected to be improved [21]:

- Capacity/capacitance
- Cyclability
- Rate capability
- Energy/power density

In addition to metal oxides, graphene/non-metal composites are also attracting great interest for being used as anode materials for lithium-ion batteries. In this respect, silicon (Si) anodes are introduced as one of the promising options for future batteries. Thanks to its potential in forming a stable alloy containing 4.4 lithium per silicon, the theoretical capacity of silicon is more than 10 times that of graphite being 4200 mA h g^{-1}. As anodes in lithium-ion batteries, silicon offers a variety of advantages over their rivals: high theoretical capacity, low discharge rate, no memory effect, environmentally friendly, and natural abundance. Moreover, silicon anodes show a voltage plateau at 370 mV against Li/Li+, which means stable operating voltage [6]. Despite all these properties that introduce silicon anodes as outstanding assets for lithium-ion energy storage, the intrinsic drawbacks of the material put obstacles in the way of its utilization in batteries. One major problem of silicon is its large expansion/contraction during lithiation/delithiation [46]. The expansion of silicon up to 400% of its original volume results in anode failure due to cracking. Cracking and loss of contact between anode and its substrate (usually copper) as well as formation of unstable interfaces with solid electrolyte affect the cyclabilty and rate performance of the anode, thus reducing the device functionality in a drastic manner. Therefore, researchers have tried to overcome this deficiency through creating nanostructured silicon anodes, which show a better response to volume fluctuations. For obtaining a better cyclability and performance, silicon composites have been synthesized [8]. In this way, different materials are explored such as Si/CNTs [47, 48, 49], Si/C [50, 51], Si/graphene [52–55], Si/polymer [56–58], and so on. Although these materials reveal significant improvement in anode functionality, there still remain issues to be solved: production cost, the complexity of the processes employed, and the use of dangerous materials during synthesis. In this scenario, silicon/RGO composites and Si/RGO/C composites with their three-dimensional structure offer improved electrical conductivity and higher stability [59]. Such structures combine the three-dimensional high theoretical capacity of silicon with flexibility and conductivity of graphene and will constitute the future batteries for portable and flexible electronic devices.

4.11 Graphene in Cathodes

Parallel to anode materials, new cathode materials for lithium-ion batteries have also been under development [9]. Basically, the same strategies (followed in anode enhancement) can be found here. In comparison to anodes, cathodes in commercially available batteries are more complex in structure. These are lithium-based compounds such as lithium cobaltite ($LiCoO_2$, layered structure), lithium nickel cobalt aluminum oxide ($LiNi_{0.8}Co_{0.15}Al_{0.05}O_2$ or NCA, layered structure), lithium nickel manganese cobalt oxide ($LiNi_{1/3}Co_{1/3}Mn_{1/3}O_2$ or NMC, layered structure), lithium manganese oxide ($LiMn_2O_4$ or LMO, spinel structure), and lithium ferro phosphate (LiFePO4 or LFP, olivine structure) [60, 61]. Each material comes with its advantages and shortages. For instance, LMO and LFP are low-cost materials but their capacity is also low. On the other hand, NMC and NCA offer a better capacity but they are expensive due to the high cost of Ni and Co.

The capacity of all these electrodes falls below 200 mA h g^{-1}, which is lower than the capacity of the negative electrode. Therefore, works on lithium-ion battery cathodes are mainly aimed at improving the electrical properties of the cathode. There is also some concern about the structural instabilities and loss of capacity and cyclability. Therefore, composites of different materials with graphene-based products have been synthesized and electrically characterized

[62, 63]. Using graphene, capacities beyond the theoretical capacity of existing cathodes are obtained.

4.12 Graphene in Other Types of Lithium Batteries

Lithium batteries come in different types including lithium-ion, lithium-polymer, lithium-sulfur, and lithium-air (the new member of the family) all of which work with lithium ion as charge carrier. Sodium-ion batteries (NIBs), on the other hand, use sodium ions as charge carriers. The advantage of NIBs over lithium-ion is that sodium salts are cheap and abundant. In addition, the current trend in the consumption of lithium in lithium-ion batteries has created a great demand for this material, which is, unfortunately, limited in abundance. However, compared to lithium-ion batteries, NIBs offer lower charge capacity (115 $mA\,h\,g^{-1}$ at 3.6 V). Due to the fact that sodium and lithium are a lot similar in a number of properties, it is expected that these batteries act as alternatives to lithium-ion batteries [64]. However, the larger size of sodium ion (1.02 Å compared to 0.76 Å of lithium ions) has a limitation on the application of these batteries since it affects the cycling process negatively [65]. Similar to lithium-ion batteries, NIBs make use of carbon as the anode material although composite materials such as $SnSe_x$ and compounds such as sodium titanate have also been introduced [66, 67]. The cathode side, on the other hand, usually is based on sodium iron phosphate or even lithium iron phosphate.

Obviously, the performance of the device depends highly on the materials and their structure. Thus, as the first and main strategy in cathode enhancement, coating cathode with conductive carbon layer and its effect on battery operation has been the focus of many researchers [68, 69]. In NIBs, special attention has been given to positive electrodes so far and a number of materials are proposed including sulphides, phosphates, fluorides, oxides, and sulphates [70].

Recently, a growing interest has been observed on using carbonaceous materials including graphene to fabricate high-capacity electrode materials for these batteries [70–77]. Whether as additive or the mainframe, graphene-based materials have proved to be efficient in improving battery performance. However, this field needs more investigation in order to obtain high-capacity devices that the world of electronics demands.

Summary

Application of graphene-based materials in advanced lithium-ion batteries was reviewed. A systematic approach was followed starting from the history of batteries to currently available types emphasizing on lithium-ion batteries as the storage devices for portable electronics as well as future hybrid and electric cars. Current challenges in developing batteries were noted and the role of graphene in improving the device performance was highlighted. The tremendous work on graphene and its derivatives in enhancing the lithium-ion battery operation and the huge interest in following the same strategy in sodium-ion devices – the new member of the family – reveals the importance of this material, which is believed to be a star in the sky of materials science and engineering.

References

1. *Annual Energy Outlook 2015 with Projections to 2040 (AEO2015)* (2015) s.l.: U.S. Energy Information Administration, April.
2. *Energy Study 2014, Reserves, Resources and Availability of Energy Resources* (2014) BGR. Hannover: s.n..
3. Frankfurt School-UNEP Centre (2015) *Global Trends in Renewable Energy Investment 2015*, Frankfurt School of Finance & Management gGmbH, Frankfurt.
4. Hsing, D., Kang, P., Chen, M., and Ogunseitan, O.A. (2013) *Potential environmental and human health impacts of rechargeable lithium batteries in electronic waste. Environ. Sci. Technol.*, **47**, 5495–5503.
5. Song, M.K., Park, S., Alamgir, F.M., Cho, J., and Liu, M. (2011) *Nanostructured electrodes for lithium-ion and lithium-air batteries: the latest developments, challenges, and perspectives. Mater. Sci. Eng. R*, **72**, 203–252.
6. Armstrong, M.J., O'Dwyer, C., Macklin, W.J., and Holmes, J.D. (2014) *Evaluating the performance of nanostructured materials as lithium-ion battery electrodes. Nano Res.*, **7**, 1–62.
7. Etacheri, V., Marom, R., Elazari, R., Salitra, G., and Aurbach, D. (2011) *Challenges in the development of advanced Li-ion batteries: a review. Energy Environ. Sci.*, **4**, 3243–3262.
8. Liu, L., Lyu, J., Li, T., and Zhao, T. (2016) *Well-constructed silicon-based materials as high-performance lithium-ion battery anodes. Nanoscale*, **8**, 701–722.
9. Zhu, N., Liu, W., Xue, M., Xie, Z., Zhao, D., Zhang, M., Chen, J., and Cao, T. (2010) *Graphene as a conductive additive to enhance the high-rate capabilities of electrospun Li4Ti5O12 for lithium-ion batteries. Electrochim. Acta*, **55**, 5813–5818.
10. *About Graphite*, http://energizerresources.com/2013-04-22-01-24-57 (accessed 22 February 2016).
11. Crowe, P. *Hybrid Cars*, http://www.hybridcars.com/graphite-for-ev-li-ion-batteries-could-be-north-america-sourced/ (accessed 22 February 2016).
12. Sur, U.K. (2012) *Graphene: a rising star on the horizon of materials science. Int. J. Electrochem.*, **2012**, 12. Article ID 237689.
13. Avouris, P. and Dimitrakopoulos, C. (2012) *Graphene synthesis and applications. Mater. Today*, **15**, 86–97.
14. Park, S. and Ruoff, R.S. (2009) *Chemical methods for the production of graphenes. Nat. Nanotechnol.*, **4**, 217–224.
15. Choi, W., Lahiri, I., Seelaboyina, R., and Kang, Y.S. (2010) *Synthesis of graphene and its applications: a review. Crit. Rev. Solid State Mater. Sci.*, **35**, 52–71.
16. Allen, M.J., Tung, V.C., and Kaner, R.B. (2010) *Honeycomb carbon: a review of graphene. Chem. Rev.*, **110**, 132–145.
17. Ovid'ko, I.A. (2013) *Mechanical properties of graphene. Rev. Adv. Mater. Sci.*, **34**, 1–11.
18. Novoselov, K.S., Fal'ko, V.I., Colombo, L., Gellert, P.R., Schwab, M.G., and Kim, K. (2012) *A roadmap for graphene. Nature*, **490**, 192–200.
19. Castro Neto, A.H., Guinea, F., Peres, N.M.R., Novoselov, K.S., and Geim, A.K. (2009) *The electronic properties of graphene. Rev. Mod. Phys.*, **81**, 109–161.
20. Novoselov, K.S., Morozov, S.V., Mohinddin, T.M.G., Ponomarenko, L.A., Elias, D.C., Yang, R., Barbolina, I.I., Blake, P., Booth, T.J., Jiang, D., Giesbers, J., Hill, E.W., and Geim, A.K. (2007)

Electronic properties of graphene. Phys. Stat. Sol. (B), **244**, 4106–4111.

21. Wu, Z.-S., Ren, W., Xu, L., Li, F., and Cheng, H.-M. (2011) *Doped graphene sheets as anode materials with superhigh rate and large capacity for lithium ion batteries. ACS Nano*, **5**, 5463–5471.
22. Wang, H., Yuan, X., Zeng, G., Wu, Y., Liu, Y., Jiang, Q., and Gu, S. (2015) *Three dimensional graphene based materials: synthesis and applications from energy storage and conversion to electrochemical sensor and environmental remediation. Adv. Colloid Interface Sci.*, **221**, 41–59.
23. Raccichini, R., Varzi, A., Passerini, S., and Scrosati, B. (2015) *The role of graphene for electrochemical energy storage. Nat. Mater.*, **14**, 271–279.
24. Zhu, Y., Murali, S., Cai, W., Li, X., Suk, J.W., Potts, J.R., and Ruoff, R.S. (2010) *Graphene and graphene oxide: synthesis, properties, and applications. Adv. Mater.*, **22**, 1–19.
25. Endo, M., Kim, C., Nishimura, K., Fujino, T., and Miyashita, K. (2000) *Recent development of carbon materials for Li ion batteries. Carbon*, **38**, 183–197.
26. Wu, Y.P., Rahm, E., and Holze, R. (2003) *Carbon anode materials for lithium ion batteries. J. Power Sources*, **114**, 228–236.
27. Luo, L., Zhao, P., Yang, H., Liu, B., Zhang, J.-G., Cui, Y., Yu, G., Zhang, S., and Wang, C.-M. (2015) *Surface coating constraint induced self-discharging of silicon nanoparticles as anodes for lithium ion batteries. Nano Lett.*, **15**, 7016–7022.
28. Morris, R.S., Dixon, B.G., Gennett, T., Raffaelle, R., and Heben, M.J. (2004) *High-energy rechargeable Li-ion battery based on carbon nanotube technology. J.Power Sources*, **138**, 277–280.
29. Bellucci, S. (ed) (2011) *Physical Properties of Ceramic and Carbon Nanoscale Structures, Lecture Notes in Nanoscale Science and Technology 11*, Springer-Verlag, Berlin.
30. Song, B., Yang, J., Zhao, J., and Fang, H. (2011) *Intercalation and diffusion of lithium ions in a carbon nanotube bundle by ab initio molecular dynamics simulations. Energy Environ. Sci.*, **4**, 1379–1384.
31. Wu, S., Xu, R., Lu, M., Ge, R., Iocozzia, J., Han, C., Jiang, B., and Lin, Z. (2015) *Graphene-containing nanomaterials for lithium-ion batteries. Adv. Energy Mater.*, 1500400 (40 pages).
32. Prosini, P.P., Carewska, M., Maroni, F., Tossici, R., and Nobili, F. (2015) *A lithium-ion battery based on LiFePO4 and silicon/reduced graphene oxide nanocomposite. Solid State Ionics*, **283**, 145–151.
33. Wang, H., Maiyalagan, T., and Wang, X. (2012) *Review on recent progress in nitrogen-doped graphene: synthesis, characterization, and its potential applications. ACS Catal.*, **2**, 781–794.
34. Quan, B., Yu, S.-H., Chung, D.Y., Jin, A., Park, J.H., Sung, Y.-E., and Piao, Y. (2014) *Single source precursor-based solvothermal synthesis of heteroatom-doped graphene and its energy storage and conversion applications. Nat. Sci. Rep.*, **4**, http://dx.doi.org/10.1038/srep05639.
35. Hu, Y., He, D., Wang, Y., Fu, M., An, X., and Zhao, X. (2016) *Defect introduced graphene sheets with hole structure as lithium-ion battery anode. Mater. Lett.*, **164**, 278–281.
36. Gómez-Navarro, C., Meyer, J.C., Sundaram, R.S., Chuvilin, A., Simon, K., Burghard, M., Kern, K., and Kaiser, U. (2010) *Atomic structure of reduced graphene oxide. Nano Lett.*, **10**, 1144–1148.
37. Fang, H., Zhao, L., Yue, W., Wang, Y., Jiang, Y., and Zhang, Y. (2015) *Facile and large-scale preparation of sandwich-structured graphene-metal oxide composites as anode materials for Li-ion batteries. Electrochim. Acta*, **186**, 397–403.
38. Zhang, L., Wang, Z., Wang, H., Yang, K., Wang, L., Li, X., Zhang, Y., and Dong, H. (2016) *Preparation and electrochemical performances of CoO/3D graphene composite as anode for lithium-ion batteries. J. Alloys Compd.*, **656**, 278–283.
39. Deng, Y., Fang, C., and Chen, G. (2016) *The developments of SnO2/graphene nanocomposites as anode materials for high performance lithium ion batteries: a review. J. Power Sources*, **304**, 81–101.

40. Xiang, J., Dong, D., Wen, F., Zhao, J., Zhang, X., Wang, L., and Liu, Z. (2016) *Microwave synthesized self-standing electrode of MoS2 nanosheets assembled on graphene foam for high-performance Li-Ion and Na-Ion batteries. J. Alloys Compd.*, **660**, 11–16.
41. Siwińska-Stefańska, K. and Kurc, B. (2015) *Preparation and application of a titanium dioxide/graphene oxide anode material for lithiumeion batteries. J. Power Sources*, **299**, 286–292.
42. Zhang, C., Zhang, P., Dai, J., Zhang, H., Xie, A., and Shen, Y. (2016) *Facile synthesis and electrochemical properties of MoO2/reduced graphene oxide hybrid for efficient anode of lithium-ion battery. Ceram. Int.*, **42**, 3618–3624.
43. Dai, J., Wang, M., Song, M., Li, P., Zhang, C., Xie, A., and Shen, Y. (2016) *A novel synthesis of ZnO/N-doped reduced graphene oxide composite as superior anode material for lithium-ion batteries. Scr. Mater.*, **112**, 67–70.
44. Yuan, G., Wang, G., Wang, H., and Bai, J. (2016) *Half-cell and full-cell investigations of 3D hierarchical MoS2/graphene composite on anode performance in lithium-ion batteries. J. Alloy. Compd.*, **660**, 62–72.
45. Wu, Z.-S., Zhou, G., Yin, L.-C., Ren, W., Li, F., and Cheng, H.-M. (2012) *Graphene/metal oxide composite electrode materials for energy storage. Nano Energy*, 107–131.
46. Sethuraman, V.A., Srinivasan, V., and Newman, J. (2013) *Analysis of electrochemical lithiation and delithiation kinetics in silicon. J. Electrochem. Soc.*, **160**, A394–A403.
47. Li, X., Zhang, G., Zhang, L., Zhong, M., and Yuan, X. (2015) *Silicon/graphite/carbon nanotubes composite as anode for lithium ion battery. Int. J. Electrochem. Sci.*, **10**, 2802–2811.
48. de las Casas, C. and Li, W. (2012) *A review of application of carbon nanotubes for lithium ion battery anode material. J. Power Sources*, **208**, 74–85.
49. Ye, M., Hu, C., Lv, L., and Qu, L. (2016) *Graphene-winged carbon nanotubes as high-performance lithium-ion batteries anode with super-long cycle life. J. Power Sources*, **305**, 106–114.
50. Badi, N., Erra, A.R., Hernandez, F.C.R., Okonkwo, A.O., and Hobosyan, M. (2014) *Low-cost carbon-silicon nanocomposite anodes for lithium ion batteries. Nanoscale Res. Lett.*, **9**, 360.
51. Zhang, Y., Zhao, Z.G., Zhang, X.G., Zhang, H.-L., Li, F., Liu, C., and Cheng, H.-M. (2008) *Pyrolytic carbon-coated silicon/carbon nanotube composites: promising application for Li-ion batteries. Int. J. Nanomanuf.*, **2**, 4–15.
52. Son, I.H., Park, J.H., Kwon, S., Park, S., Rümmeli, M.H., Bachmatiuk, A., Song, H.J., Ku, J., Choi, J.W., Choi, J.-M., Doo, S.-G., and Chang, H. (2015) *Silicon carbide-free graphene growth on silicon for lithium-ion battery with high volumetric energy density. Nat. Commun.*, **6**.
53. Sun, W., Hu, R., Zhang, H., Wang, Y., Yang, L., Liu, J., and Zhu, M. (2016) *A long-life nano-silicon anode for lithium ion batteries: supporting of graphene nanosheets exfoliated from expanded graphite by plasma-assisted milling. Electrochim. Acta*, **187**, 1–10.
54. Luo, Z., Xiao, Q., Lei, G., Li, Z., and Tang, C. (2016) *Si nanoparticles/graphene composite membrane for high performance silicon anode in lithium ion batteries. Carbon*, **98**, 373–380.
55. Feng, J., Zhang, Z., Ci, L., Zhai, W., Ai, Q., and Xiong, S. (2015) *Chemical dealloying synthesis of porous silicon anchored by in situ generated graphene sheets as anode material for lithium-ion batteries. J. Power Sources*, **287**, 177–183.
56. Guo, J. and Wang, C. (2010) *A polymer scaffold binder structure for high capacity silicon anode of lithium-ion battery. Chem. Commun.*, **46**, 1428–1430.
57. Koo, B., Kim, H., Cho, Y., Lee, K.T., Choi, N.-S., and Cho, J. (2012) *A highly cross-linked polymeric binder for high-performance silicon negative electrodes in lithium ion batteries. Angewandte Chemie*, **51**, 8762–8767.
58. Chan, C.K., Patel, R.N., O'Connell, M.J., Korgel, B.A., and Cui, Y. (2010) *Solution-grown silicon nanowires for lithium-ion battery anodes. ACS Nano*, **4**, 1443–1450.

59. Feng, K., Ahn, W., Lui, G., Park, H.W., Kashkooli, A.G., Jiang, G., Wang, X., Xiao, X., and Chen, Z. (2016) *Implementing an in-situ carbon network in Si/reduced graphene oxide for high performance lithium-ion battery anodes. Nano Energy*, **19**, 187–197.
60. Brodd, R.J. (ed) (2013) *Batteries for Sustainability: Selected Entries from the Encyclopedia of Sustainability Science and Technology*, Springer-Verlag, New York.
61. Choi, M., Kim, H.-S., Lee, Y.M., Choi, W.-K., and Jin, B.-S. (2016) *The high electrochemical performance of Li3V2(PO4)3 supported by graphene and carbon-nanofibers for advanced Li-ion batteries. Mater. Res. Bull.*, **73**, 211–218.
62. Kucinskis, G., Bajars, G., and Kleperis, J. (2013) *Graphene in lithium ion battery cathode materials: a review. J. Power Sources*, **240**, 66–79.
63. Arbizzani, C., Da Col, L., De Giorgio, F., Mastragostino, M., and Soavi, F. (2015) *The role of modified graphene in cathode formulations for lithium-ion batteries. ECS Trans.*, **66**, 139–147.
64. Rui, X., Sun, W., Wu, C., Yu, Y., and Yan, Q. (2015) *An advanced sodium-ion battery composed of carbon coated Na3V2 (PO4) 3 in a porous graphene network. Adv. Mater.*, **27**, 6670–6676.
65. Van Nghia, N., Ou, P.-W., and Hung, I.-M. (2015) *Synthesis and electrochemical properties of sodium manganese-based oxide cathode material for sodium-ion batteries. Electrochim. Acta*, **161**, 63–71.
66. Zhang, Z., Zhao, X., and Li, J. (2016) *SnSex flower like composites as anode materials for sodium ion batteries. Mater. Lett.*, **162**, 169–172.
67. Zhang, Y., Hou, H., Yang, X., Chen, J., Jing, M., Wu, Z., Jia, X., and Ji, X. (2016) *Sodium titanate cuboid as advanced anode material for sodium ion batteries. J. Power Sources*, **305**, 200–208.
68. Wang, H., Jiang, D., Zhang, Y., Li, G., Lan, X., Zhong, H., Zhang, Z., and Jiang, Y. (2015) *Self-combustion synthesis of Na3V2(PO4)3 nanoparticles coated with carbon shell as cathode materials for sodium-ion batteries. Electrochim. Acta*, **155**, 23–28.
69. Cui, D., Chen, S., Han, C., Ai, C., and Yuan, L. (2016) *Carbothermal reduction synthesis of carbon coated Na2FePO4F for lithium ion batteries. J. Power Sources*, **301**, 87–92.
70. Balogun, M.-S., Luo, Y., Qiu, W., Liu, P., and Tong, Y. (2016) *A review of carbon materials and their composites with alloy metals for sodium ion battery anodes. Carbon*, **98**, 162–178.
71. Ruan, Y.-L., Wang, K., Song, S.-D., Han, X., and Cheng, B.-W. (2015) *Graphene modified sodium vanadium fluorophosphate as a high voltage cathode material for sodium ion batteries. Electrochim. Acta*, **160**, 330–336.
72. Liu, Y., Cheng, Z., Sun, H., Arandiyan, H., Li, J., and Ahmad, M. (2015) *Mesoporous Co3O4 sheets/3D graphene networks nanohybrids for high-performance sodium-ion battery anode. J. Power Sources*, **273**, 878–884.
73. Yin, L., Wang, Y., Han, C., Kang, Y.-M., Ma, X., Xie, H., and Wu, M. (2016) *Self-assembly of disordered hard carbon/graphene hybrid for sodium-ion batteries. J. Power Sources*, **305**, 156–160.
74. Ramos, A., Cameán, I., Cuesta, N., Antuña, C., and García, A.B. (2016) *Expanded graphitic materials prepared from micro- and nanometric precursors as anodes for sodium-ion batteries. Electrochim. Acta*, **187**, 496–507.
75. Darwiche, A., Dugas, R., Fraisse, B., and Monconduit, L. (2016) *Reinstating lead for high-loaded efficient negative electrode for rechargeable sodium-ion battery. J. Power Sources*, **304**, 1–8.
76. Zhang, X., Hu, Y., Zhu, D., Xie, A., and Shen, Y. (2016) *A novel porous CuO nanorod/rGO composite as a high stability anode material for lithium-ion batteries. Ceram. Int.*, **42**, 1833–1839.
77. Xiao, L., Cao, Y., Henderson, W.A., Sushko, M.L., Shao, Y., Xiao, J., Wangc, W., Engelhard, M.H., Nie, Z., and Liu, J. (2016) *Hard carbon nanoparticles as high-capacity, high-stability anodic materials for Na-ion batteries. Nano Energy*, **19**, 279–288.

5
Graphene-Based Membranes for Separation Engineering

Luisa M. Pastrana-Martínez, Sergio Morales-Torres, José L. Figueiredo, and Adrián M.T. Silva

5.1
Introduction

The ideal efficient membrane should be as thin as possible to maximize permeance, mechanically robust to resist the applied pressure, and present a well-defined pore size distribution to increase selectivity.

Graphene, because of its extraordinary properties, such as atomic thickness, high mechanical strength, large surface area, two-dimensional structure, and large Young modulus [1], has been considered to be perfect membrane material, because it is only atomically thick and thus may be made into membranes that minimize transport resistance and maximize permeate flux [2–9]. Theoretical models predict that graphene would be an ideal membrane in terms of selectivity and permeability by creating sub-nanometer sized pores in the two-dimensional lattice exhibiting superior separation performances for gas and water molecules [4, 5, 10–14].

Graphene oxide (GO), one of the most interesting graphene derivatives, has recently claimed large attention for its potential use in membrane science because of its cost-effective, facile and large-scale production [15–19]. The two-dimensional structure and tunable physicochemical properties of GO offer an exciting opportunity to make novel types of membranes by stacking GO nanosheets [7, 20, 21]. In the GO membrane, the pores or channels are generated by tailoring the interplanar spacing between two-dimensional GO sheets. The strong hydrogen bonds within the individual GO sheets hold the sheets together to form a lamellar structure. This offers several straightforward strategies to control the pore size in the tiny channels of GO via tuning the type and amount of oxygen functional groups.

During the past years, graphene was believed to be impermeable to gases, vapor, and liquids. However, recent studies showed breakthrough discoveries in the permeation properties of GO membranes. Nair *et al.* [7] reported that water vapor permeates through the membranes, but not other molecules, such as H_2, N_2, Ar, ethanol, hexane, acetone, decane, and propanol. Kim *et al.* [2] also reported highly permeable and selective GO membranes for separating mixtures

Graphene Technology: From Laboratory to Fabrication,
First Edition. Edited by Soroush Nazarpour and Stephen R. Waite.

of gases of industrial relevance, whereas Joshi *et al.* [3] described the permeation of aqueous solutions containing ions and neutral molecules through GO, reporting "ultrafast" transport properties. Therefore, graphene-based materials can act as promising candidates for preparing high-quality membranes for water and gas technologies, opening outstanding opportunities in separation science.

This chapter focuses on the preparation, properties, and applications of graphene-based membranes, considering gas separation and water treatment (purification and desalination) and their potential use as next-generation membranes in the oil and gas industry. The chapter offers a comprehensive discussion of graphene-based membranes, from a broad perspective of separation applications up to advanced membranes with improved separation performance.

5.2 Preparation of Graphene-Based Membranes

The perfect graphene sheet exhibits complete barrier properties to all molecules and ions [22, 23]. This is because the electron density of the graphene aromatic rings is enough to repel atoms and molecules trying to pass through them. However, graphene can be produced selectively permeable by creating pores in the impermeable two-dimensional lattice, so it would act as an efficient selective sieve membrane [12, 24–31]. The immense potential of porous, single-layer graphene membranes is therefore built on the ability to introduce controlled subnanometer-sized pores at high density over large areas of graphene, and to understand the relationship between pore structures and the resulting transport properties [12]. The most important methods to create pores in graphene include electron beam [24–26], ultraviolet-induced oxidative etching [27, 28], exposure to high-temperature atmospheric oxygen [29], hydrogen plasma [30], and ion irradiation [12, 31], as shown in Figure 5.1.

Graphene derivatives, such as GO, are mostly considered for membranes fabrication and for use in gas separation and water purification because of mass production, controllable pore size distribution, and application under high pressure conditions [17, 19]. GO is commonly obtained from an aqueous dispersion of graphite oxide. The first method to produce graphite oxide was reported in the nineteenth century by Brodie [32]. This method was modified by several investigators including Staudenmaier [33] and Hummers and Offeman [34]. Graphite oxide can be exfoliated into oxidized carbon monolayers (GO) by sonication in different solvents such as, water, alcohols, and certain organic solvents due to the ionization of oxygen surface groups attached to the graphene layers [35–37]. The structure of GO is still a subject of debate. A popular structural model was proposed by Lerf and Klinowski [38, 39] in which GO sheets exhibited epoxide and hydroxyl groups, mostly at the basal planes of the sheets, as well as carboxylic acid groups at the edges (Figure 5.2).

GO can then be reassembled into a laminate structure as a solid membrane with robust mechanical strength as well as flexibility [1, 40, 41]. Because of the presence

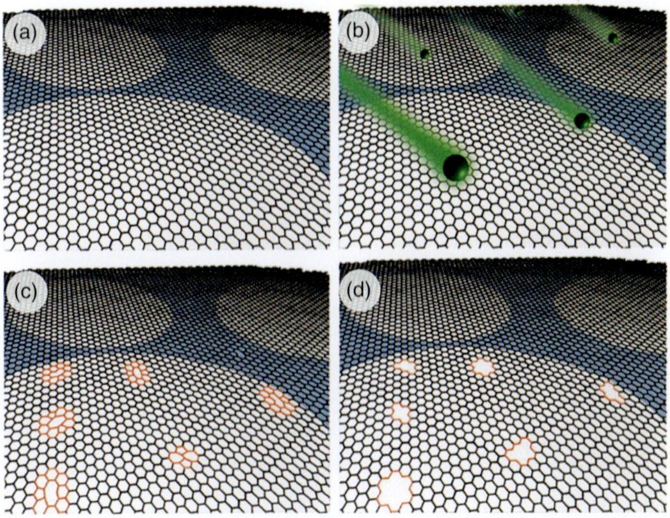

Figure 5.1 Processes to create controlled pores in a graphene membrane: (a–c) gallium ion bombardment of graphene sheets; (d) etching with oxidizing solution. (Reproduced from Ref. [12] with permission from the American Chemical Society, Copyright 2014.)

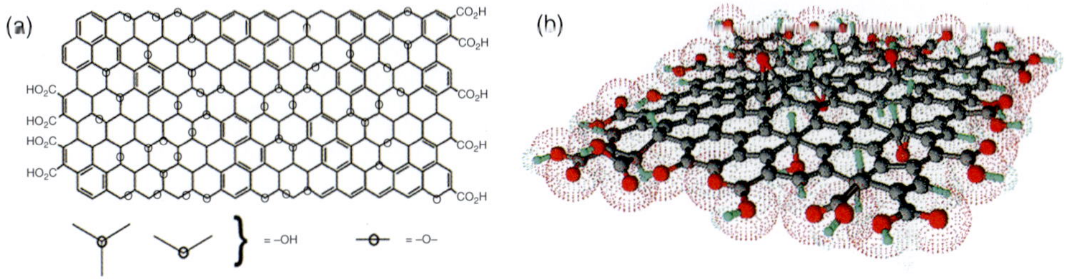

Figure 5.2 (a) Lerf–Klinowski structural model for GO showing carboxylic acid groups at the edges, and phenol hydroxyl and epoxide groups mainly at the basal planes. (Reproduced from Ref. [38] with permission from the American Chemical Society, Copyright 1998.) (b) 3D view of a GO sheet. (Reproduced from Ref. [37] with permission from the American Chemical Society, Copyright 2009.)

of oxygen functional groups and adsorbed water molecules on both sides of the basal planes, the interlayer space of the lamellar GO membrane (*d*-distance) is significantly larger than that of a multilayer stack of graphene and it depends on the preparation method and the amount of intercalated water molecules [1, 15]. The filtration mechanism in GO membranes is still unclear, mainly due to the complexity in both microstructures and spatial distribution of chemical functional groups in the GO membrane.

As previously mentioned, GO membranes were impermeable to liquids and gases, while unimpeded permeation of water vapor was observed [7]. Wei *et al.* [42] reported water permeation in GO membranes using atomistic simulations

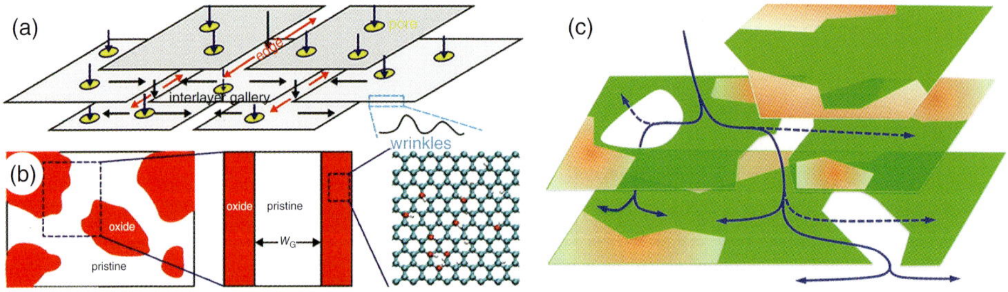

Figure 5.3 Microstructures of graphene-derived membranes: (a) The percolated water transport channel is composed of interlayer, interedge spaces, wrinkles, and pores within the graphene sheets. (b) The pristine and oxidized patterns on GO (left) are modeled in a quasi-2D molecular model (center) with oxygen-containing functionalized groups on both sides (right) (Reprinted from Ref. [42] with permission from the American Chemical Society, Copyright 2014). (c) Sketches of water motion through the stacked GO layers with channels (capillaries) in the vicinity of the edges of graphene sheets. The oxidized area is denoted by red color and graphene without oxidation by green. Solid (dotted) blue lines are (un)favorable paths for water permeation. All edges are assumed to be passivated by hydrophilic edge groups. (Reprinted from Ref. [43] with permission from the American Chemical Society. Copyright, 2013.)

and theoretical analysis. The authors found that the fast flow of water across graphene derivative membranes may be attributed mainly to their porous microstructures (e.g., expanded interlayer gallery, wide channels formed at wrinkles, holes, and inter-edge spaces), followed by the less significant enhancement by boundary slip (Figure 5.3a–b). Boukhvalov *et al.* reported that water molecules can permeate through the interconnected nanochannels formed between GO nanosheets and follow a tortuous path primarily over the hydrophobic non-oxidized surface rather than the hydrophilic oxidized region of GO [43, 44]. Both the capillary driven force and low-friction flow confined between the non-oxidized GO facilitate the extremely fast flow of water molecules (Figure 5.3c) [7]. Huang *et al.* [45] reported that nanostrand-channeled GO ultrafiltration membranes with a narrow pore size distribution (3–5 nm) feature superior separation performance without sacrificing the rejection rate compared to unmodified GO membranes, because of their highly porous structure and significantly reduced channel length.

For ion sieving, size exclusion appears to be the dominant mechanism. Ions smaller in size than the GO nanochannel can permeate through the GO membrane at a speed orders of magnitude faster than would occur by simple diffusion [3]. Ideally, GO membranes should have ~0.3 nm of interspacing between sheets [7]. However, when the GO membrane is immersed into an ionic solution, the spacing of the nanosheets increases up to ~0.9 nm. In this case, any ionic species with a hydrated radius larger than 0.45 nm is blocked, while those with smaller radius could enter through the nanochannel (Figure 5.4a) [3].

By adjusting the GO spacing through appropriately sized spacers between GO layers, a broad spectrum of GO membranes can be made, each one being able to

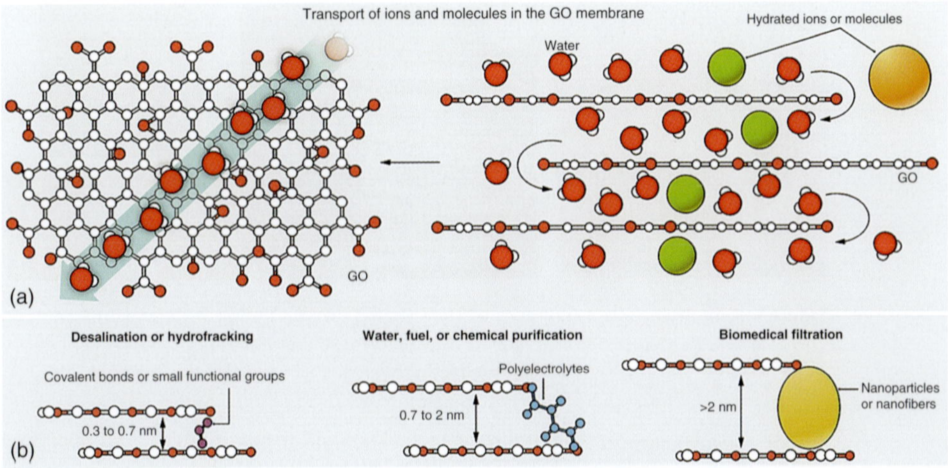

Figure 5.4 GO membranes. (a) Water and small-sized ions and molecules (compared with the void spacing between stacked GO nanosheets) permeate superfast in the GO membrane, but larger species are blocked. (b) The separation capability of the GO membrane is tunable by adjusting the nanochannel size. (Reprinted from Ref. [44] with permission from The American Association for the Advancement of Science, Copyright 2014.)

precisely separate target ions and molecules within a specific size range from bulk solution. In this way, GO membranes may be ideally tailored for applications in water purification, wastewater reuse, and pharmaceutical and fuel separation, or even for biomedical applications [44] (Figure 5.4b).

GO membranes can be fabricated by different methods, including spin/spray-coating, drop casting, Langmuir–Blodgett (L–B), vacuum filtration or layer-by-layer (LbL) assembly, among others.

The first methodology consists in the spin/spray-coating of a smooth substrate (i.e., copper foil) using a GO suspension [2, 7, 46]. Nair *et al.* [7] described a procedure to fabricate freestanding GO membranes of ~1 cm in diameter after etching away a central part of the copper foil with a HNO_3 treatment. Finally, the membranes were cleaned with deionized water and dried on a hot plate (<50 °C). Dense stacking occurs because the face-to-face attractive capillary forces created by the spray- or spin-coating can overcome the repulsive forces between the edges of GO sheets. Therefore, the initial deposition is mainly governed by capillary interactions between the faces of GO sheets, rather than by the electrostatic forces between the GO edges (Figure 5.5).

GO dispersions can be also re-assembled into membranes with controlled thickness through a drop-casting process. This methodology induced the nanocapillary network formation for selective ion penetration and water purification properties of freestanding GO membranes [47, 48]. In this method, the GO colloidal suspension is drop-casted onto a substrate with a smooth surface, such as silica or paper, and dried at room temperature. Once deposited, the van der Waals

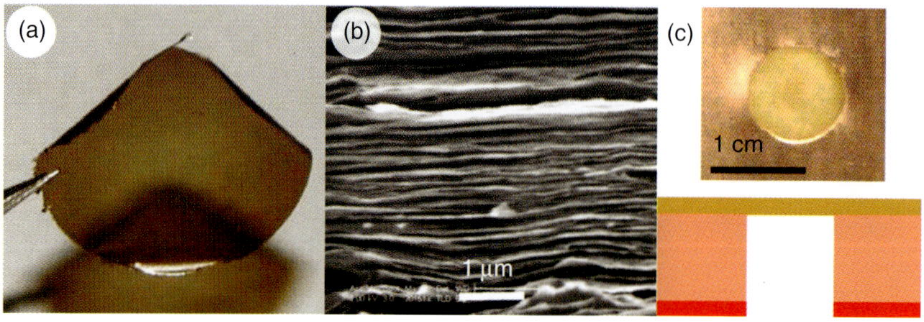

Figure 5.5 (a) Photo of a 1-mm-thick GO film peeled off of a Cu foil. (b) TEM of the film cross-section (c) mounted on Cu hole for separation test. (Reprinted from Ref. [7] with permission from The American Association for the Advancement of Science, Copyright 2012.)

forces are sufficient to keep the GO sheets strongly adhered to the substrate. The oxygen functional groups, located asymmetrically in the GO membrane, maintain a certain distance during the drop-casting process, creating void spaces between non-oxidized regions forming a series of nanocapillaries within the film.

The preparation of GO membranes by L–B assembly includes the dispersion of GO in a volatile organic solvent (i.e., methanol or chloroform) [37, 49] and then, spreading it onto the water surface. As the solvent evaporates, the molecules are trapped on the water surface, forming a monolayer. The film can be transferred to a solid substrate (e.g., by dip-coating), forming a monolayer coating over a large area. Bi- and tri-layers of GO films can be prepared by drying the as-formed films in an oven, and then depositing another layer of GO under the same conditions. The number of layers and thickness of GO membranes can be readily controlled using the L–B technique.

Vacuum filtration has been frequently used for the fabrication of self-standing GO membranes for gas separation and water treatment [1–3, 50, 51]. The GO suspension is filtered through the supports leading to the deposition of the GO membrane on the filter. The thickness of the GO membrane is controlled by adjusting the volume of the GO suspension. The most commonly used filters for preparing membranes are anodic aluminum oxide (AAO) discs and polymeric membranes (i.e., mixed cellulose ester membranes). The AAO discs are preferred due to their rigidity, which can promote a tighter packing of the sheets under vacuum filtration, leading to stronger membranes [3, 52].

Membranes fabricated by vacuum filtration have interactions between the nanosheets, such as electrostatic repulsion, van der Waals attractive forces and hydrogen bonding. Compression induced by vacuum sucking was responsible for aligning the GO sheets perpendicularly to the flow direction and narrowing the inter-spacing between the oriented GO sheets [53]. It has been reported [51] that the filtration rate affects the pore structure substantially and, consequently, the separation capacity of the resulting membranes, since it affects the way the GO sheets are stacked on the surface of the porous substrate. Slow filtration leads to smaller thickness of the GO stacks and thus the space between the aggregated stacks is of smaller size.

It is noteworthy that pristine GO membranes made via solution filtration demonstrate outstanding mechanical durability in dry conditions [3]; however, the dispersion of GO membranes in water (i.e., low stability) has been also reported due to the electrostatic repulsion between the GO sheets (negatively charged upon hydration) that can overcome van der Waals attraction or hydrogen bonding that holds the sheets together [20, 21, 54]. Thus, such free-standing membranes made by simple filtration methods are not likely to survive the cross-flow testing conditions, typical of real-world membrane operations [49].

One way to solve this issue consists in forming stable bonds between GO nanosheets, stabilizing the membrane integrity. These cross-linking spacers could also act as barriers to form a new channel, which would greatly enhance water permeability. To this end, few studies have focused on fabricating cross-linked GO membranes in an effort to improve their mechanical integrity [20, 52, 55–59] during LbL assembly. The LbL method is ideal to introduce an interlayer stabilizing force via covalent bonding, electrostatic interaction, or both effects during layer deposition [44]. The GO membrane thickness can be readily controlled by varying the number of LbL deposition cycles. This technique is highly scalable for membranes, as opposed with the synthesis of monolayer graphene membranes, which requires the manufacturing of large-sized graphene sheets and the punching of nanopores with a narrow size distribution [60].

Yeh *et al.* [52] also reported that pristine GO membranes can disintegrate in water, but they can become stable if they are cross linked by multivalent cationic metals. They found that metal cations, such as Al^{3+} (released from an AAO filter disc during the filtration) or Mn^{2+} (a by-product from GO synthesis), can act as cross-linkers in the sheets and strengthen the final membrane. Significant enhancement in mechanical stiffness and fracture strength of GO membranes can be also achieved by the addition of a small amount (<1 wt%) of Mg^{2+} and Ca^{2+} during the filtration [55]. These results can be readily rationalized in terms of the chemical interactions between the functional groups of the GO sheets and the divalent metals ions. Hu *et al.* [56] reported GO membranes by LbL assembling with positively charged polyelectrolytes such as poly(allylamine hydrochloride) (PAH) via electrostatic interaction. Although the binding forces between assembled polyelectrolyte layers are mainly contributed by electrostatic interaction, hydrophobic interaction and hydrogen bonding may also be involved to a lower extent.

Hu and Mi [20] prepared a water filtration membrane with layered GO nanosheets on a polysulfone support coated with polydopamine and 1,3,5-benzenetricarbonyl trichloride (TMC) as cross-linkers (Figure 5.6). The ionic flux of sulfonated polyethersulfone (SPES) membranes can be improved up to 19% by the incorporation of 10% GO as compared to unmodified SPES membrane. The strong interfacial interactions due to GO nanofillers into the SPES matrix improve the thermal and mechanical properties of the nanocomposite membranes [57]. An *et al* [58]. studied GO membranes covalently cross-linked with borate inspired by cell wall assembly of higher-order plants. The study has reported a remarkable enhancement by over 255% in the stiffness of the graphene film by

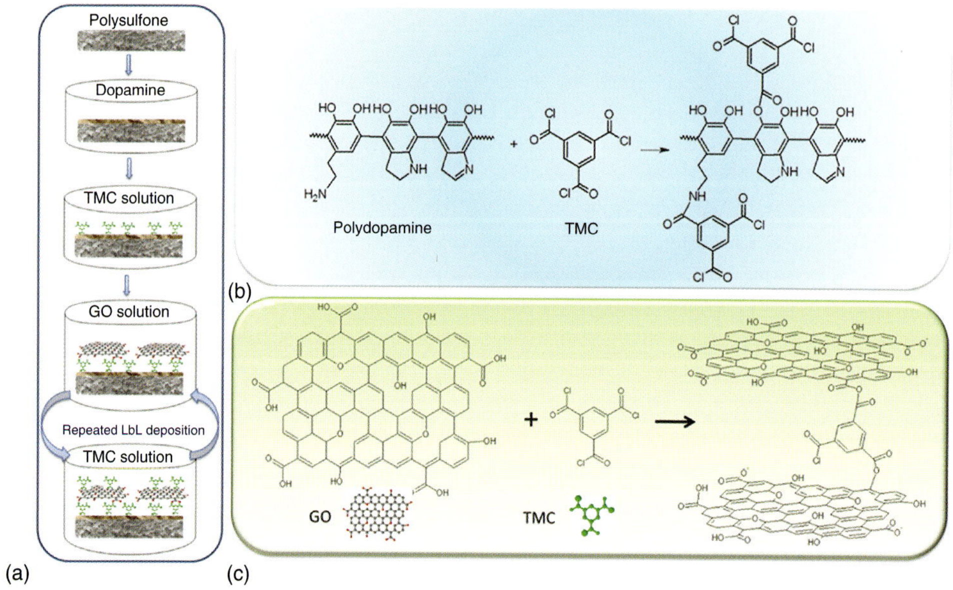

Figure 5.6 Schematic illustration of (a) an LbL procedure to synthesize the GO membrane; (b) the mechanism of reactions between polydopamine and TMC, and (c) the mechanism of reactions between GO and TMC. (Reprinted from Ref. [20] with permission from the American Chemical Society, Copyright 2013.)

adding 0.94 wt% boron to the final composite GO film. The strong cross-linking between adjacent nanosheets has been attributed to the formation of covalent bonds between borate ions and the hydroxyl groups on the GO nanosheets.

5.3 Graphene-based Membranes for Separation Applications

Graphene-based membranes have shown promising performance in gas separation and water treatment due to their unique "size sieving" effect. Both nanoscaled interlayer spacing between two individual flakes and selective structural defects on the flakes have been claimed to be responsible for the observed separation performance.

5.3.1 Gas Separation

Theoretical studies have investigated the permeability and selectivity of graphene-based membranes toward different gases [6, 10, 61–63].

In 2012, Nair *et al.* [7] tested the permeance of several gases, including He, H_2, N_2, and Ar through self-standing GO membranes with a thickness ranging from 0.1 to 10 mm. The leakage rate was monitored by the variations in the inner pressure of the sealed containers over a period of several days. The results revealed that no noticeable reduction in pressure was observed for any tested gas, including He, H_2, N_2, and Ar, suggesting that the GO membranes were completely impermeable to these tested gases, which is similar to that previously mentioned for graphene monolayers. This work was the first report showing that GO membranes may separate different molecules through interlayer spacing between GO sheets.

Koening *et al.* [27] demonstrated the experimental evidence of gas separation through individual porous graphene membranes. Pores were introduced in graphene by ultraviolet-induced oxidative etching and the molecular transport through them was measured using a pressurized blister test as well as mechanical resonance. The deformation of the film surface due to gas pressure on the feed side was measured to quantify the permeability, with a higher deflection indicating a lower permeability due to the gas molecules not being able to pass the graphene barrier. The etched porous graphene membrane was found to be selective for H_2 and CO_2 over larger molecules such as Ar, N_2, and CH_4, suggesting the presence of pores slightly larger than 0.3 nm in size (Figure 5.7). This work demonstrated the great potential of using nanoporous graphene as a promising membrane material for gas separation by molecular sieving.

Kim *et al.* [2] demonstrated highly permeable and selective thin (i.e., few-layered) defective graphene and GO membranes for separating mixtures of gases of industrial relevance. They reported that graphene itself displays outstanding gas separation properties, with O_2/N_2 separation properties above the so-called upper bound [64] (Figure 5.8). Further studies on GO membranes showed outstanding properties in terms of CO_2 selectivity (e.g., CO_2/H_2, CO_2/N_2, and CO_2/CH_4). The authors reported that the gas-transport properties are quite sensitive to the preparation method and the presence of water in the GO structures,

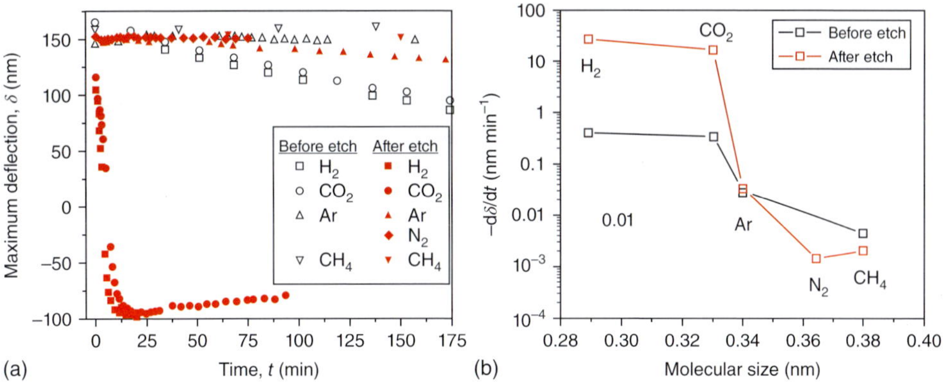

Figure 5.7 Comparison of gas permeation rates of pristine and porous graphene membranes: (a) maximum deflection of the membrane surface versus time before and after etching; (b) gas permeance versus molecular size, revealing selective permeation of H_2 and CO_2 over larger gas molecules after etching. The connecting lines show the measurements before (black) and after (red) etching. (Reprinted from Ref. [27] with permission from Nature Publishing Group, Copyright 2012.)

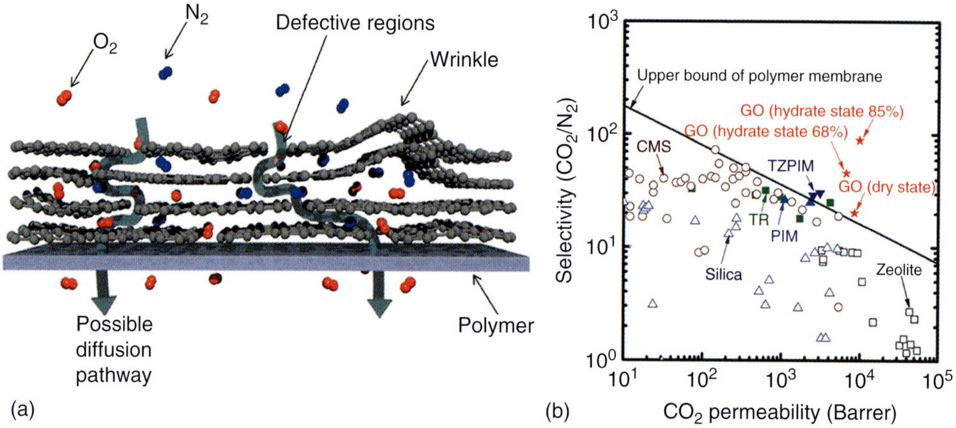

Figure 5.8 Permeation of gas molecules through laminates of GO membranes. (a) Permeation of gases through defect regions. (b) Performance of GO membranes relative to other high-performance polymeric (TR, thermally rearranged polymer; TZPIM, tetrazole functional polymer of intrinsic microporosity; PIM, polymer of intrinsic microporosity) and inorganic (CMS, carbon molecular sieve, zeolite, silica) membranes. (Reprinted from Ref. [2] with permission from The American Association for the Advancement of Science, Copyright 2013.)

as these variables change the organization of the laminate structure, thereby changing the shapes and sizes of the openings available for gas transport [65].

Li *et al.* [50] prepared GO membranes with a thickness of about 1.8 nm by a facile filtration process. The membranes showed mixture separation selectivities as high as 3400 and 900 for H_2/CO_2 and H_2/N_2 mixtures, respectively, through selective structural defects on GO. They suggested that the major transport pathway in few nanometer thick GO films relied mostly on certain structural defects present along the GO sheets rather than due to interlayer spacing.

Our group [51] fabricated a series of GO membranes by varying the filtration rate of the starting GO suspension as well as the amount of the surface functional groups of the employed GO. The as-produced GO membranes were robust enough for incorporation in various membrane modules as well as to test their performance in gas permeability studies. The results showed that the slow filtration derived membranes were almost impermeable to *m*-xylene vapor and exhibited a good separation performance for several gas pairs (H_2/N_2, H_2/CO, H_2/CH_4, H_2/C_2H_6, H_2/C_4H_{10}, and H_2/SF_6). Gas transport through the void space between GO stacks, or discontinuities of GO layers, was dominant over transport through the inter-layer space.

5.3.2 Water Treatment

Graphene-based membranes have been developed and applied for water applications, including treatment, purification, and even desalination.

Nair *et al.* [7] were the first authors that reported one of the most exciting and revolutionary physical properties of GO membranes as ideal filters for water, opening outstanding opportunities in separation science. They found that GO membranes can be impermeable to all molecules tested except to water vapor (Figure 5.9). Water molecules can flow through the empty space between the layers in a GO laminate (interlayer *d*-distance) because this space is what is required for one layer of water molecules. When the other molecules that were tested tried to enter such space, they found that the graphene capillaries were shrunk, or clogged, with water molecules. In addition, the permeation of water through the GO membrane was as fast as the evaporation through an open aperture (unimpeded).

Cohen-Tanugi *et al.* [4] showed the high performance of monolayer graphene to effectively filter salt (NaCl) from water, using classical molecular dynamics. They report the desalination performance of such membranes as a function of pore size, chemical functionalization, and pressure. In another work [66], the authors reported the mechanical strength of nanoporous graphene as a reverse osmosis (RO) membrane using molecular dynamics simulation. They showed that an appropriate substrate with openings smaller than 1 μm would allow nanoporous graphene to withstand pressures exceeding 570 bar or 10 times more than typical pressures for seawater RO. Moreover, nanoporous graphene membranes can maintain their ultrahigh permeability even at low pressures (<100 bar), indicating that they are promising membranes for water desalination [67].

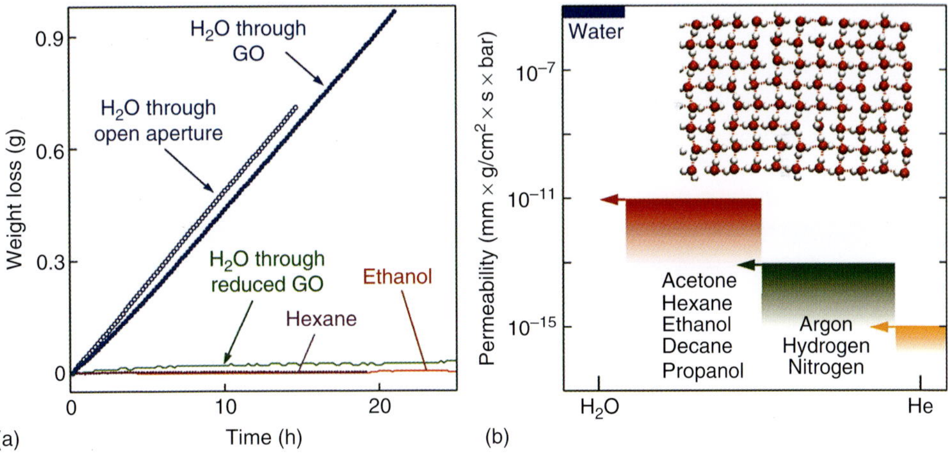

Figure 5.9 Permeation through GO membrane. (a) Weight loss for a container sealed with a GO film. No loss was detected for ethanol, hexane, and so on, but water evaporated from the container as freely as through an open aperture (blue curves). (b) Permeability of GO membrane with respect to water and various small molecules (arrows indicate the upper limits set by our experiments). (Inset) Schematic representation of the structure of monolayer water inside a graphene capillary with $d = 7$ Å. (Reprinted from Ref. [7] with permission from The American Association for the Advancement of Science, Copyright 2012.)

O'Hern *et al.* [12] reported selective ionic transport through controlled subnanometer pores (created by chemical oxidation of nucleated defects) in graphene membranes. The results revealed that the created pores were selective for cations at short oxidation times, consistent with electrostatic repulsion from negatively charged functional groups terminating the pore edges. At longer oxidation times, the pores allowed transport of salt but prevented the transport of a larger organic molecule, indicative of steric-size exclusion.

In a separate study, Qiu *et al.* [68] reported an approach to create corrugation on reduced graphene oxide (rGO) flakes by hydrothermal treatment of the rGO dispersions. The researchers revealed that the amplitude of GO corrugation can be simply controlled by hydrothermal treatment temperature, and the corrugation could form nanochannels in the rGO membranes. Colloids of Au and Pt nanoparticles with average diameters of ~13 and ~3 nm, respectively, were used for a series of filtration tests. As a result, the 90-rGO (rGO hydrothermally treated at 90 °C) membrane showed no permeation for both Au and Pt, indicating a channel size smaller than 3 nm. When rGO was treated at higher temperatures (100 and 120 °C), the membranes allowed only Pt to pass through, but at a still higher temperature (150 °C) both Au and Pt nanoparticles could pass through the membrane.

The rejection of ions and organic dye molecules by GO membranes has also been intensively studied. The selectivity of this process is mainly attributed to both size exclusion and interactions (including chemical and electrostatic ones) with functional groups. Hu *et al.* [20] reported that GO membranes display a relatively low rejection of monovalent and divalent salts (6–46%) and moderate/high rejection of organic dyes (46–66% and 93–95% for methylene blue and rhodamine B, respectively). Similar results were also reported by Han *et al.* [21] with large performance for the retention of organic dyes (especially for the charged dyes), which was ascribed to the mechanism of physical sieving and electrostatic interaction.

Huang *et al.* [54] investigated the effects of pH, salt concentration, and pressure on the controlled separation performance of GO membranes. The ionic strength and protonation (deprotonation) of the carboxylic acid groups on the GO sheets were responsible for the variation in the inter-spacing between GO sheets and the pore size of nanochannels, leading to the corresponding separation performance. At low pH values and high salt concentrations, the repulsion forces between the negatively charged GO sheets dramatically decrease and shrink the space between them, resulting in the reduction of the flux and increase in the rejection rate. Furthermore, the collapsed nanochannels at high pressure could be recovered by releasing the applied pressure, due to their unique elastic properties.

Sun *et al.* [47] demonstrated selective ion penetration properties of freestanding GO membranes; they reported fast permeation of sodium salts, slow permeation of heavy-metal salts, and no permeation of Cu ions and organic contaminants such as rhodamine B. The nanocapillaries formed within the GO membranes might be responsible for the permeation of metal cations and the strong coordination between transition-metal ions, and the oxygen-containing functional groups might give rise to the much slower permeation of heavy metal ions through GO membranes compared to sodium salts (Figure 5.10). The authors also reported the

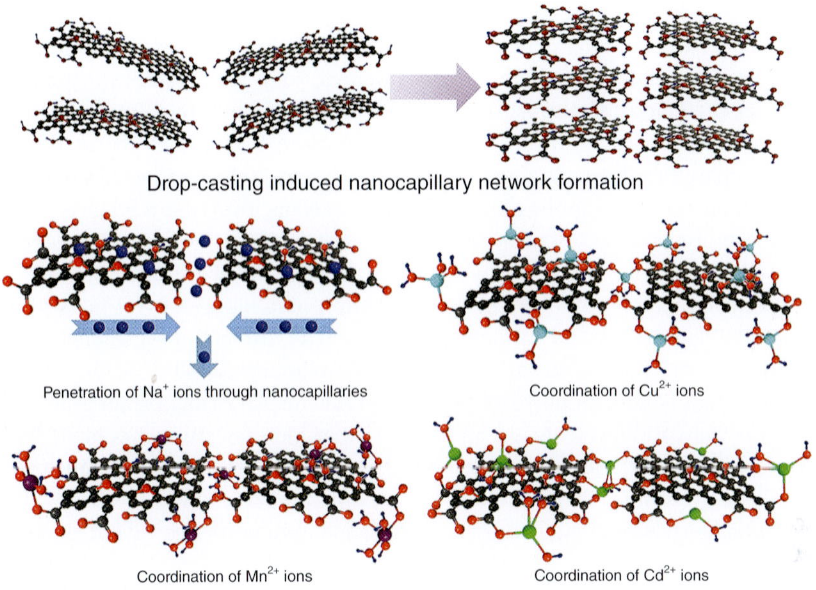

Figure 5.10 Schematic diagrams of a GO membrane and the interaction with different ions. (Reprinted from Ref. [47] with permission from the American Chemical Society, Copyright 2013.)

selective trans-membrane transport properties of alkali and alkaline earth cations (Na^+, K^+, Mg^{2+}, Ca^{2+}, and Ba^{2+}) through a membrane composed of stacked and overlapped GO sheets. The simulation of the transmembrane transport of cations indicated that the coordinative interactions are responsible for selectivity [48]. The authors [48] also demonstrated the efficient recovery of acids from iron-based electrolytes using GO membranes. The trans-membrane transport of H^+ was two orders of magnitude greater than that of Fe^{3+}. Notably, when the concentration of $FeCl_3$ sources was reduced to some extent (e.g., 0.01 mol l^{-1}), the Fe^{3+} cations could be blocked by GO membranes. The mechanism for the effective separation of H^+ from Fe^{3+} was discussed, indicating that the molecular sieving effect of GO nanocapillaries and the coordination between Fe^{3+} and GO were responsible for the effective capture of Fe^{3+} while the rapid propagation of protons through the hydrogen-bonding networks along the water layers formed within the interlayer spacing was responsible for the fast migration of H^+.

Joshi *et al.* [3] described the permeation of aqueous solutions of ions and neutral molecules through GO membranes, reporting "ultrafast" transport properties. They focused on the transport of ions (e.g., NaCl, $MgCl_2$, and $K_3[Fe(CN)_6]$) as well as water-soluble neutral molecules (glycerol and sucrose). The authors observed a remarkably sharp cut-off in the permeation properties as a function of the hydration size of the permeating solute. Small hydrated ions, ranging in size from K^+ to Mg^{2+}, had similar permeation properties, whereas larger species were essentially prohibited to permeate through the GO membrane. They estimated that the membrane was essentially impermeable to molecules with a hydration radius larger than approximately 4.5 Å. The authors also reported very high partition coefficients for salts with GO membranes, suggesting values for the salt partition coefficient, K (the ratio of the salt concentration in the GO membrane to that in the external solution), as high as 10 or more. Such values are extraordinary compared to those obtained for typical polymers, which are widely used in desalination and other applications involving salt transport (Figure 5.11).

The use of membranes consisting of polymeric matrices and GO is an alternative that combines the processability of polymers and the unique properties of GO materials for water treatment. The hydrophilic nature of GO makes it a good candidate to improve antimicrobial properties, increase permeability, and enhance the mechanical strength of polymeric membranes [69–73]. GO membranes cross-linked with polydopamine and TMC displayed significant rejection of monovalent and divalent salts and moderate or high rejection of organic dyes [20]. Choi *et al.* [59] created an assembly of GO nanosheets on a polyamide (PA) thin-film composite (TFC) membrane for desalination driven by RO (Figure 5.12). The GO material was coated onto the PA-TFC membrane surface via LbL deposition of oppositely charged nanosheets, which resulted in increased hydrophilicity and reduced surface roughness of the RO membrane. The altered properties of the RO membrane due to the GO nanosheets resulted in enhanced resistance to protein fouling and increased chlorine resistance. Wang *et al.* [74] used rGO in polyethersulfone (PES) membranes to fabricate TFC membranes, which were tested in forward osmosis (FO) of salty water. TFC membranes modified with rGO presented higher water fluxes than those prepared with the neat PES support,

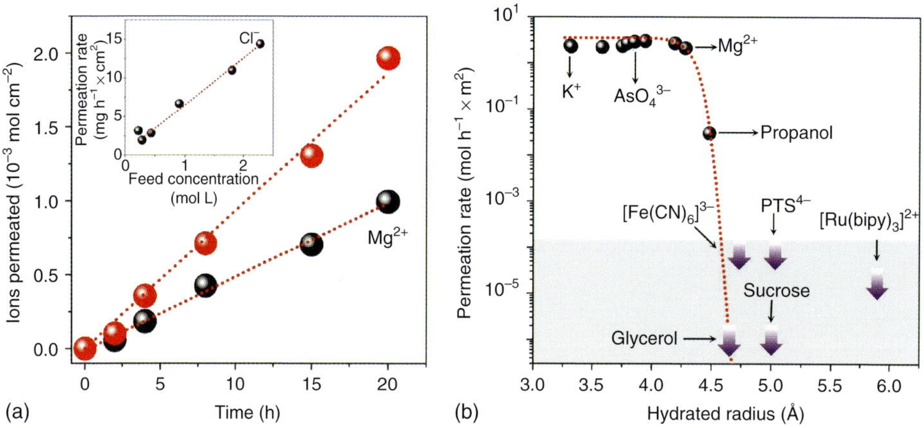

Figure 5.11 (a) Ion permeation through GO laminates. Permeation through a 5-mm-thick GO membrane from the feed compartment with a 0.2 M solution of $MgCl_2$. (Inset) Permeation rates as a function of C in the feed solution; (b) Sieving through atomic-scale mesh. The shown permeation rates are normalized per 1 M feed solution and are measured by using 5-mm-thick membranes. (Reprinted from Ref. [3] with permission from The American Association for the Advancement of Science, Copyright 2014.)

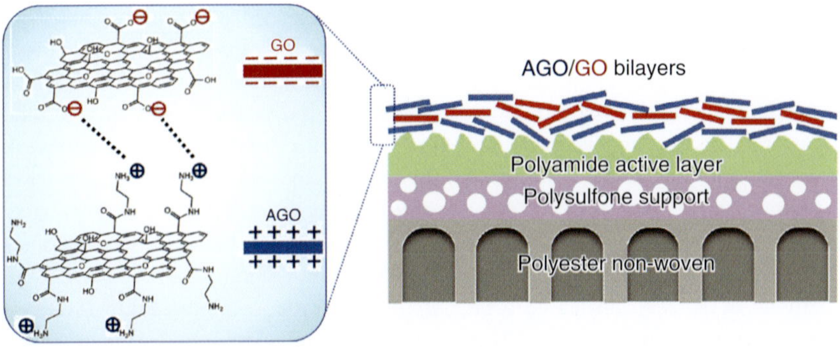

Figure 5.12 Schematic illustration of a multilayered GO coating on a PA thin-film composite membrane surface via LbL deposition of oppositely charged GO and aminated-GO (AGO) nanosheets. (Reprinted from Ref. [59] with permission from the American Chemical Society, Copyright 2013.)

the enhanced performance being due to a low structural parameter of the resulting membrane and the reduction of the internal concentration polarization. TFC membranes with GO (0.12 wt%) also improved the pure water flux upto 80% compared to the neat TFC membrane, without significantly affecting salt selectivity in RO experiments [75].

Similar findings were reported concerning GO–TiO_2 (GOT) nanocomposite membranes for water treatment [76, 77] (Figure 5.13) and GO-based polyelectrolyte membranes assembled to synthesize the cross-linked GO membrane for water reuse [20, 58]. These cross-linking spacers could also act as barriers to form new channels that would greatly enhance water permeability.

In view of the versatile mass transport characteristics of GO membranes, focus on their pervaporation separation behavior has also been reported.

Figure 5.13 Cross-sectional M-GO/GOT SEM images at different magnifications. (a) 1000×; (b) 2500× (inset corresponds to the freestanding GO membrane: M-GO). (Reprinted from Ref. [77] with permission from Elsevier, Copyright 2015.)

Tang *et al.* [78] used the pervaporation process for dehydration of ethanol using free-standing GO membranes assembled by a pressurized ultrafiltration method. Experimental results suggested that the interlayer spacing was determined by both the packing density of GO nanosheets and the water content in the feed solution. The packing density was sensitively affected by the ultrafiltration pressure applied during membrane formation. By tuning the ultrafiltration pressure, a high separation performance with water permeability of 13 800 Barrer (1 Barrer = 3.348×10^{-19} kmol m m^{-2} s^{-1} Pa^{-1}) and water/ethanol selectivity of 227 was achieved for dehydration of a 85 wt% ethanol aqueous solution at 24 °C.

Huang *et al.* [79] fabricated a GO membrane prepared by vacuum filtration onto a ceramic hollow fiber. This GO membrane exhibited excellent water permeation of dimethyl carbonate/water mixtures through a pervaporation process. The permeate water content reached 95.2 wt% and a high permeation flux of 1702 g m^{-2} h^{-1} at 25 °C and 2.6 wt% feed water content.

Tsou *et al.* [80] synthesized GO/polyacrylonitrile (mPAN) membranes prepared by a pressure-, vacuum-, and evaporation-assisted self-assembly technique. The membranes were applied to dehydrate 1-butanol mixtures by pervaporation. The results showed that the membranes exhibited exceptional pervaporation performance at 30 °C (permeate water of 99.6 wt% and permeation flux of 2.54 kg m^{-2} h^{-1}); moreover, the membrane sustained its operating stability at a relatively high temperature of 70 °C and a high water concentration of 99.5 wt% was maintained with a permeation flux of 4.34 kg m^{-2} h^{-1}.

Currently, the mentioned problems associated with nanoporous graphene membranes are expected to be overcome, and the commercialization of this type of membranes already started; Perforene™ (from Lockheed Martin Corporation, USA) being an example. The Perforene membrane was developed by placing holes that are ≤1 nm of size in a graphene membrane. These holes are small enough to trap the ions while dramatically improving the flow-through of water molecules, reducing clogging and pressure on the membrane. The Perforene material works by removing sodium, chlorine, and other ions from seawater and other sources. Furthermore, the graphene-based membranes are seen as being 500× thinner and 99% more energy efficient than those employed in RO [81].

Graphene is also being explored for its use in various stages of oil and gas processes. However, the difficulty of dispersing large graphene flakes in aqueous media creates problems in water-based muds [82]. Instead of graphene, GO provides a more stable material for aqueous dispersions and it maintains the sheet-like morphology that would allow the desired pore-plugging through filter cake formation. Kosynkin *et al.* [83] showed that adding platelets of GO to a common water-based drilling fluid decreased the losses of the fluid to the surrounding rock, as compared to a standard mixture of clays and polymers used in the drilling industry today. By methylating the GO through an esterification reaction, the stability of GO in saline environments is increased, making it an ideal candidate for the next generation of fluid-loss-control additives.

Another work has been patented [84], which is related to the recovery of oil from subterranean reservoirs and, more particularly, to new and improved secondary recovery operations using flood water including a dispersion of a hydrophilic graphite oxide for mobility control [85]. McCoy *et al.* [86] have revealed the possibility of GO in stabilizing crude oil in water emulsion. At highly acidic pH values, fully reversible flocculation of emulsion droplets can be achieved, whereas when adjusted to high pH, the flocculation is irreversible, which can be interpreted as a permanent chemical change in GO. The interfacial charge of the GO and oil–water interface is the over-riding drive for the exceptional stability of acidic GO emulsions. Nguyen *et al.* [87] reported the use of GO as a thermal stability additive for polymer solutions in polymer-enhanced oil recovery.

Baker Hughes Incorporated, an oilfield service company, patented the use of graphene as a component in fluids used for retrieving oil and gas [88], since the graphene can infiltrate the rock and act as a seal, preventing leakage that is more fluid. On the other hand, Lockheed Martin Corp. is testing nanofilters using its patented Perforene (graphene sheets with precisely sized holes as small as 1 nm) for water remediation in the oil and gas industry. The company states that the ultimate goal is water desalination, but more feasible and immediate uses can be found in the oil and gas industry, in particular to clean drilling wastewaters, where the requirements in terms of the quality of the graphene and hole sizes are less challenging. This goal only requires 50–100 nm sized holes, compared to the 1 nm holes needed for desalination [89].

5.4 Conclusions

In summary, membranes based on graphene and graphene derivatives are standing up as promising candidates for precise and selective gas separation and water purification processes due to their high permeance, outstanding mechanical properties, and facile fabrication.

The structure and preparation of graphene-based membranes such as nanoporous graphene membranes, free-standing GO membranes and chemically cross-linked GO membranes, among others, have been briefly summarized. It has been demonstrated that graphene-based membranes can exhibit mechanical strength, high permeance, and excellent selectivity.

Graphene-based membranes show excellent selectivity and superior performance for multiple gas mixtures as well as for water/wastewater treatment. However, several challenges still exist in their path to industrial application, including the ability needed to synthesize large quantities of graphene material with enough mechanical strength to be stable under higher hydrostatic pressures, as well as the precise design essential of GO membranes, including the pattern of oxidized regions, the interlayer space, the pore size, and the number of graphene layers. These studies are expected to attract a great deal of interest from the industry, including the oil and gas industry, due to the importance of separation processes in their sustainability.

Acknowledgments

This work was financially supported by Project POCI-01-0145-FEDER-006984 – Associated Laboratory LSRE-LCM funded by FEDER funds through COMPETE2020 - Programa Operacional Competitividade e Internacionalização (POCI) – and by national funds through FCT - Fundação para a Ciência e a Tecnologia. LMPM and AMTS acknowledge the FCT Investigator program (IF/01248/2014 and IF/01501/2013, respectively), with financing from the European Social Fund and the Human Potential Operational Program. SMT acknowledges financial support from FCT grant SFRH/BPD/108981/2015.

References

1. Dikin, D.A., Stankovich, S., Zimney, E.J., Piner, R.D., Dommett, G.H.B., Evmenenko, G., Nguyen, S.T., and Ruoff, R.S. (2007) Preparation and characterization of graphene oxide paper. *Nature*, **448** (7152), 457–460.
2. Kim, H.W., Yoon, H.W., Yoon, S.-M., Yoo, B.M., Ahn, B.K., Cho, Y.H., Shin, H.J., Yang, H., Paik, U., Kwon, S., Choi, J.-Y., and Park, H.B. (2013) Selective gas transport through few-layered graphene and graphene oxide membranes. *Science*, **342** (6154), 91–95.
3. Joshi, R.K., Carbone, P., Wang, F.C., Kravets, V.G., Su, Y., Grigorieva, I.V., Wu, H.A., Geim, A.K., and Nair, R.R. (2014) Precise and ultrafast molecular sieving through graphene oxide membranes. *Science*, **343** (6172), 752–754.
4. Cohen-Tanugi, D. and Grossman, J.C. (2012) Water desalination across nanoporous graphene. *Nano Lett.*, **12** (7), 3602–3608.
5. Konatham, D., Yu, J., Ho, T.A., and Striolo, A. (2013) Simulation insights for graphene-based water desalination membranes. *Langmuir*, **29** (38), 11884–11897.
6. Du, H., Li, J., Zhang, J., Su, G., Li, X., and Zhao, Y. (2011) Separation of hydrogen and nitrogen gases with porous graphene membrane. *J. Phys. Chem. C*, **115** (47), 23261–23266.
7. Nair, R.R., Wu, H.A., Jayaram, P.N., Grigorieva, I.V., and Geim, A.K. (2012) Unimpeded permeation of water through helium-leak–tight graphene-based membranes. *Science*, **335** (6067), 442–444.
8. Aghigh, A., Alizadeh, V., Wong, H.Y., Islam, M.S., Amin, N., and Zaman, M. (2015) Recent advances in utilization of graphene for filtration and desalination of water: a review. *Desalination*, **365**, 389–397.
9. Mahmoud, K.A., Mansoor, B., Mansour, A., and Khraisheh, M. (2015) Functional graphene nanosheets: the next generation membranes for water desalination. *Desalination*, **356**, 208–225.
10. Jiang, D.-E., Cooper, V.R., and Dai, S. (2009) Porous graphene as the ultimate membrane for gas separation. *Nano Lett.*, **9** (12), 4019–4024.
11. Suk, M.E. and Aluru, N.R. (2010) Water transport through ultrathin graphene. *J. Phys. Chem. Lett.*, **1** (10), 1590–1594.
12. O'Hern, S.C., Boutilier, M.S.H., Idrobo, J.-C., Song, Y., Kong, J., Laoui, T., Atieh, M., and Karnik, R. (2014) Selective ionic transport through tunable subnanometer pores in single-layer graphene membranes. *Nano Lett.*, **14** (3), 1234–1241.
13. Sint, K., Wang, B., and Král, P. (2008) Selective ion passage through functionalized graphene nanopores. *J. Am. Chem. Soc.*, **130** (49), 16448–16449.
14. Sun, C., Boutilier, M.S.H., Au, H., Poesio, P., Bai, B., Karnik, R., and Hadjiconstantinou, N.G. (2014) Mechanisms of molecular permeation through nanoporous graphene membranes. *Langmuir*, **30** (2), 675–682.
15. Zhu, Y., Murali, S., Cai, W., Li, X., Suk, J.W., Potts, J.R., and Ruoff, R.S. (2010) Graphene and graphene oxide: synthesis,

properties, and applications. *Adv. Mater.*, **22** (35), 3906–3924.
16. Eda, G. and Chhowalla, M. (2010) Chemically derived graphene oxide: towards large-area thin-film electronics and optoelectronics. *Adv. Mater.*, **22** (22), 2392–2415.
17. Huang, H., Ying, Y., and Peng, X. (2014) Graphene oxide nanosheet: an emerging star material for novel separation membranes. *J. Mater. Chem. A*, **2** (34), 13772–13782.
18. Xu, Q., Xu, H., Chen, J., Lv, Y., Dong, C., and Sreeprasad, T.S. (2015) Graphene and graphene oxide: advanced membranes for gas separation and water purification. *Inorg. Chem. Front.*, **2** (5), 417–424.
19. Hegab, H.M. and Zou, L. (2015) Graphene oxide-assisted membranes: fabrication and potential applications in desalination and water purification. *J. Membr. Sci.*, **484**, 95–106.
20. Hu, M. and Mi, B. (2013) Enabling graphene oxide nanosheets as water separation membranes. *Environ. Sci. Technol.*, **47** (8), 3715–3723.
21. Han, Y., Xu, Z., and Gao, C. (2013) Ultrathin graphene nanofiltration membrane for water purification. *Adv. Funct. Mater.*, **23** (29), 3693–3700.
22. Bunch, J.S., Verbridge, S.S., Alden, J.S., van der Zande, A.M., Parpia, J.M., Craighead, H.G., and McEuen, P.L. (2008) Impermeable atomic membranes from graphene sheets. *Nano Lett.*, **8** (8), 2458–2462.
23. Leenaerts, O., Partoens, B., and Peeters, F.M. (2008) Graphene: a perfect nanoballoon. *Appl. Phys. Lett.*, **93** (19), 193107.
24. Garaj, S., Liu, S., Golovchenko, J.A., and Branton, D. (2013) Molecule-hugging graphene nanopores. *Proc. Natl. Acad. Sci. U S A*, **110** (30), 12192–12196.
25. Merchant, C.A., Healy, K., Wanunu, M., Ray, V., Peterman, N., Bartel, J., Fischbein, M.D., Venta, K., Luo, Z., Johnson, A.T.C., and Drndić, M. (2010) DNA translocation through graphene nanopores. *Nano Lett.*, **10** (8), 2915–2921.
26. Schneider, G.F., Kowalczyk, S.W., Calado, V.E., Pandraud, G., Zandbergen, H.W., Vandersypen, L.M.K., and Dekker, C. (2010) DNA translocation through graphene nanopores. *Nano Lett.*, **10** (8), 3163–3167.
27. Koenig, S.P., Wang, L., Pellegrino, J., and Bunch, J.S. (2012) Selective molecular sieving through porous graphene. *Nat. Nanotechnol.*, **7** (11), 728–732.
28. Cheng, Y.C., Kaloni, T.P., Zhu, Z.Y., and Schwingenschlögl, U. (2012) Oxidation of graphene in ozone under ultraviolet light. *Appl. Phys. Lett.*, **101** (7), 073110.
29. Liu, L., Ryu, S., Tomasik, M.R., Stolyarova, E., Jung, N., Hybertsen, M.S., Steigerwald, M.L., Brus, L.E., and Flynn, G.W. (2008) Graphene oxidation: thickness-dependent etching and strong chemical doping. *Nano Lett.*, **8** (7), 1965–1970.
30. Wu, S., Yang, R., Shi, D., and Zhang, G. (2012) Identification of structural defects in graphitic materials by gas-phase anisotropic etching. *Nanoscale*, **4** (6), 2005–2009.
31. Russo, C.J. and Golovchenko, J.A. (2012) Atom-by-atom nucleation and growth of graphene nanopores. *Proc. Natl. Acad. Sci.*, **109** (16), 5953–5957.
32. Brodie, B.C. (1859) On the atomic weight of graphite. *Philos. Trans. R. Soc. Lond.*, **149**, 249–259.
33. Staudenmaier, L. (1898) Verfahren zur Darstellung der Graphitsäure. *Ber. Dtsch. Chem. Ges.*, **31** (2), 1481–1487.
34. Hummers, W.S. and Offeman, R.E. (1958) Preparation of graphitic oxide. *J. Am. Chem. Soc.*, **80** (6), 1339–1339.
35. Stankovich, S., Piner, R.D., Chen, X., Wu, N., Nguyen, S.T., and Ruoff, R.S. (2006) Stable aqueous dispersions of graphitic nanoplatelets via the reduction of exfoliated graphite oxide in the presence of poly(sodium 4-styrenesulfonate). *J. Mater. Chem.*, **16** (2), 155–158.
36. Paredes, J.I., Villar-Rodil, S., Martínez-Alonso, A., and Tascón, J.M.D. (2008) Graphene oxide dispersions in organic solvents. *Langmuir*, **24** (19), 10560–10564.
37. Cote, L.J., Kim, F., and Huang, J. (2009) Langmuir–Blodgett assembly of graphite oxide single layers. *J. Am. Chem. Soc.*, **131** (3), 1043–1049.
38. Lerf, A., He, H., Forster, M., and Klinowski, J. (1998) Structure of graphite

oxide revisited. *J. Phys. Chem. B*, **102** (23), 4477–4482.

39. Dreyer, D.R., Park, S., Bielawski, C.W., and Ruoff, R.S. (2010) The chemistry of graphene oxide. *Chem. Soc. Rev.*, **39** (1), 228–240.
40. Chen, C., Yang, Q.-H., Yang, Y., Lv, W., Wen, Y., Hou, P.-X., Wang, M., and Cheng, H.-M. (2009) Self-assembled free-standing graphite oxide membrane. *Adv. Mater.*, **21** (29), 3007–3011.
41. Chen, H., Müller, M.B., Gilmore, K.J., Wallace, G.G., and Li, D. (2008) Mechanically strong, electrically conductive, and biocompatible graphene paper. *Adv. Mater.*, **20** (18), 3557–3561.
42. Wei, N., Peng, X., and Xu, Z. (2014) Understanding water permeation in graphene oxide membranes. *ACS Appl. Mater. Interfaces*, **6** (8), 5877–5883.
43. Boukhvalov, D.W., Katsnelson, M.I., and Son, Y.-W. (2013) Origin of anomalous water permeation through graphene oxide membrane. *Nano Lett.*, **13** (8), 3930–3935.
44. Mi, B. (2014) Graphene oxide membranes for ionic and molecular sieving. *Science*, **343** (6172), 740–742.
45. Huang, H., Song, Z., Wei, N., Shi, L., Mao, Y., Ying, Y., Sun, L., Xu, Z., and Peng, X. (2013) Ultrafast viscous water flow through nanostrand-channelled graphene oxide membranes. *Nat. Commun.*, **4**, 2979.
46. Pham, V.H., Cuong, T.V., Hur, S.H., Shin, E.W., Kim, J.S., Chung, J.S., and Kim, E.J. (2010) Fast and simple fabrication of a large transparent chemically-converted graphene film by spray-coating. *Carbon*, **48** (7), 1945–1951.
47. Sun, P., Zhu, M., Wang, K., Zhong, M., Wei, J., Wu, D., Xu, Z., and Zhu, H. (2013) Selective ion penetration of graphene oxide membranes. *ACS Nano*, **7** (1), 428–437.
48. Sun, P., Wang, K., Wei, J., Zhong, M., Wu, D., and Zhu, H. (2014) Effective recovery of acids from iron-based electrolytes using graphene oxide membrane filters. *J. Mater. Chem. A*, **2** (21), 7734–7737.
49. Wang, X., Bai, H., and Shi, G. (2011) Size fractionation of graphene oxide sheets by pH-assisted selective sedimentation. *J. Am. Chem. Soc.*, **133** (16), 6338–6342.
50. Li, H., Song, Z., Zhang, X., Huang, Y., Li, S., Mao, Y., Ploehn, H.J., Bao, Y., and Yu, M. (2013) Ultrathin, molecular-sieving graphene oxide membranes for selective hydrogen separation. *Science*, **342** (6154), 95–98.
51. Romanos, G., Pastrana-Martínez, L.M., Tsoufis, T., Athanasekou, C., Galata, E., Katsaros, F., Favvas, E., Beltsios, K.G., Siranidi, E., Falaras, P., Psycharis, V., and Silva, A.M.T. (2015) A facile approach for the development of fine-tuned self-standing graphene oxide membranes and their gas and vapor separation performance. *J. Membr. Sci.*, **493**, 734–747.
52. Yeh, C.-N., Raidongia, K., Shao, J., Yang, Q.-H., and Huang, J. (2015) On the origin of the stability of graphene oxide membranes in water. *Nat. Chem.*, **7** (2), 166–170.
53. Putz, K.W., Compton, O.C., Segar, C., An, Z., Nguyen, S.T., and Brinson, L.C. (2011) Evolution of order during vacuum-assisted self-assembly of graphene oxide paper and associated polymer nanocomposites. *ACS Nano*, **5** (8), 6601–6609.
54. Huang, H., Mao, Y., Ying, Y., Liu, Y., Sun, L., and Peng, X. (2013) Salt concentration, pH and pressure controlled separation of small molecules through lamellar graphene oxide membranes. *Chem. Commun.*, **49** (53), 5963–5965.
55. Park, S., Lee, K.-S., Bozoklu, G., Cai, W., Nguyen, S.T., and Ruoff, R.S. (2008) Graphene oxide papers modified by divalent ions—enhancing mechanical properties via chemical cross-linking. *ACS Nano*, **2** (3), 572–578.
56. Hu, M. and Mi, B. (2014) Layer-by-layer assembly of graphene oxide membranes via electrostatic interaction. *J. Membr. Sci.*, **469**, 80–87.
57. Gahlot, S., Sharma, P.P., Gupta, H., Kulshrestha, V., and Jha, P.K. (2014) Preparation of graphene oxide nanocomposite ion-exchange membranes for desalination application. *RSC Adv.*, **4** (47), 24662–24670.
58. An, Z., Compton, O.C., Putz, K.W., Brinson, L.C., and Nguyen, S.T. (2011)

Bio-inspired borate cross-linking in ultra-stiff graphene oxide thin films. *Adv. Mater.*, **23** (33), 3842–3846.

59. Choi, W., Choi, J., Bang, J., and Lee, J.-H. (2013) Layer-by-layer assembly of graphene oxide nanosheets on polyamide membranes for durable reverse-osmosis applications. *ACS Appl. Mater. Interfaces*, **5** (23), 12510–12519.

60. Wang, E.N. and Karnik, R. (2012) Water desalination: graphene cleans up water. *Nat. Nanotechnol.*, 7 (9), 552–554.

61. Schrier, J. (2010) Helium separation using porous graphene membranes. *J. Phys. Chem. Lett.*, **1** (15), 2284–2287.

62. Hauser, A.W. and Schwerdtfeger, P. (2012) Nanoporous graphene membranes for efficient 3He/4He separation. *J. Phys. Chem. Lett.*, **3** (2), 209–213.

63. Trinh, T.T., Vlugt, T.J.H., Hägg, M.-B., Bedeaux, D., and Kjelstrup, S. (2013) Selectivity and self-diffusion of CO_2 and H_2 in a mixture on a graphite surface. *Front. Chem.*, **1**, 38.

64. Sanders, D.F., Smith, Z.P., Guo, R., Robeson, L.M., McGrath, J.E., Paul, D.R., and Freeman, B.D. (2013) Energy-efficient polymeric gas separation membranes for a sustainable future: a review. *Polymer*, **54** (18), 4729–4761.

65. Kim, H.W., Yoon, H.W., Yoo, B.M., Park, J.S., Gleason, K.L., Freeman, B.D., and Park, H.B. (2014) High-performance CO_2-philic graphene oxide membranes under wet-conditions. *Chem. Commun.*, **50** (88), 13563–13566.

66. Cohen-Tanugi, D. and Grossman, J.C. (2014) Mechanical strength of nanoporous graphene as a desalination membrane. *Nano Lett.*, **14** (11), 6171–6178.

67. Cohen-Tanugi, D. and Grossman, J.C. (2014) Water permeability of nanoporous graphene at realistic pressures for reverse osmosis desalination. *J. Chem. Phys.*, **141** (7), 074704.

68. Qiu, L., Zhang, X., Yang, W., Wang, Y., Simon, G.P., and Li, D. (2011) Controllable corrugation of chemically converted graphene sheets in water and potential application for nanofiltration. *Chem. Commun.*, **47** (20), 5810–5812.

69. Yu, L., Zhang, Y., Zhang, B., Liu, J., Zhang, H., and Song, C. (2013) Preparation and characterization of HPEI-GO/PES ultrafiltration membrane with antifouling and antibacterial properties. *J. Membr. Sci.*, **447**, 452–462.

70. Lee, J., Chae, H.-R., Won, Y.J., Lee, K., Lee, C.-H., Lee, H.H., Kim, I.-C., and Lee, J.-m. (2013) Graphene oxide nanoplatelets composite membrane with hydrophilic and antifouling properties for wastewater treatment. *J. Membr. Sci.*, **448**, 223–230.

71. Ganesh, B.M., Isloor, A.M., and Ismail, A.F. (2013) Enhanced hydrophilicity and salt rejection study of graphene oxide-polysulfone mixed matrix membrane. *Desalination*, **313**, 199–207.

72. Zhao, C., Xu, X., Chen, J., and Yang, F. (2013) Effect of graphene oxide concentration on the morphologies and antifouling properties of PVDF ultrafiltration membranes. *J. Environ. Chem. Eng.*, **1** (3), 349–354.

73. Liang, B., Zhan, W., Qi, G., Lin, S., Nan, Q., Liu, Y., Cao, B., and Pan, K. (2015) High performance graphene oxide/polyacrylonitrile composite pervaporation membranes for desalination applications. *J. Mater. Chem. A*, **3** (9), 5140–5147.

74. Wang, Y., Ou, R., Wang, H., and Xu, T. (2015) Graphene oxide modified graphitic carbon nitride as a modifier for thin film composite forward osmosis membrane. *J. Membr. Sci.*, **475**, 281–289.

75. He, L., Dumée, L.F., Feng, C., Velleman, L., Reis, R., She, F., Gao, W., and Kong, L. (2015) Promoted water transport across graphene oxide–poly(amide) thin film composite membranes and their antibacterial activity. *Desalination*, **365**, 126–135.

76. Xu, C., Cui, A., Xu, Y., and Fu, X. (2013) Graphene oxide–TiO_2 composite filtration membranes and their potential application for water purification. *Carbon*, **62**, 465–471.

77. Pastrana-Martínez, L.M., Morales-Torres, S., Figueiredo, J.L., Faria, J.L., and Silva, A.M.T. (2015) Graphene oxide based ultrafiltration membranes for photocatalytic degradation of organic pollutants in salty water. *Water Res.*, **77**, 179–190.

78. Tang, Y.P., Paul, D.R., and Chung, T.S. (2014) Free-standing graphene oxide

thin films assembled by a pressurized ultrafiltration method for dehydration of ethanol. *J. Membr. Sci.*, **458**, 199–208.
79. Huang, K., Liu, G., Lou, Y., Dong, Z., Shen, J., and Jin, W. (2014) A graphene oxide membrane with highly selective molecular separation of aqueous organic solution. *Angew. Chem. Int. Ed.*, **53** (27), 6929–6932.
80. Tsou, C.-H., An, Q.-F., Lo, S.-C., De Guzman, M., Hung, W.-S., Hu, C.-C., Lee, K.-R., and Lai, J.-Y. (2015) Effect of microstructure of graphene oxide fabricated through different self-assembly techniques on 1-butanol dehydration. *J. Membr. Sci.*, **477**, 93–100.
81. Wood, J.A., Schwartz, R., Lill, P.J. (2014) Dryer and water recovery/purification unit employing graphene oxide or perforated graphene monolayer membranes. WO2014176541A1.
82. Luo, J., Cote, L.J., Tung, V.C., Tan, A.T.L., Goins, P.E., Wu, J., and Huang, J. (2010) Graphene oxide nanocolloids. *J. Am. Chem. Soc.*, **132** (50), 17667–17669.
83. Kosynkin, D.V., Ceriotti, G., Wilson, K.C., Lomeda, J.R., Scorsone, J.T., Patel, A.D., Friedheim, J.E., and Tour, J.M. (2012) Graphene oxide as a high-performance fluid-loss-control additive in water-based drilling fluids. *ACS Appl. Mater. Interfaces*, **4** (1), 222–227.
84. Rodewald, P.G. (1976) *Oil Recovery by Waterflooding Employing Graphite Oxide for Mobility Control*, http://www.google.com.ar/patents/US3998270 (accessed 19 February 2016).
85. Rodewald, P.G. (1976) Oil recovery by waterflooding employing graphite oxide for mobility control. US3998270.
86. McCoy, T.M., Pottage, M.J., and Tabor, R.F. (2014) Graphene oxide-stabilized oil-in-water emulsions: pH-controlled dispersion and flocculation. *J. Phys. Chem. C*, **118** (8), 4529–4535.
87. Ba Dung, N., Trung Kien, N., Truong Han, B., Duy Khanh, P., Xuan Loc, D., and Phuong Tung, N. (2015) The impact of graphene oxide particles on viscosity stabilization for diluted polymer solutions using in enhanced oil recovery at HTHP offshore reservoirs. *Adv. Nat. Sci.: Nanosci. Nanotechnol.*, **6** (1), 015012.
88. Monteiro, O. and Quintero, L. (2012) Graphene-containing fluids for oil and gas exploration and production. WO2012129302A2.
89. http://www.graphene-info.com/lockheed-martin-tests-nanofilters-oil-and-gas-wastewater-management (accessed February 2015).

6
Graphene Coatings for the Corrosion Protection of Base Metals

Robert V. Dennis, Nathan A. Fleer, Rachel D. Davidson, and Sarbajit Banerjee

6.1
Introduction to Corrosion

The unfortunate and relentless problem of corrosion of base metals has plagued mankind for thousands of years. In industrialized societies, the massive costs of maintaining and repairing infrastructure can be attributed to a large measure to the weathering and corrosion of structural components [1–4]. The complexity of corrosion processes makes estimation of the true costs of addressing such degradation phenomena rather difficult; however, several studies have attempted to provide some accounting of the financial impact of corrosion on the economy. An influential report from the United States Federal Highway Administration, published in 2002, estimated that the direct cost due to corrosion to the United States economy in 1998 was USD276 billion (amounting to as much as 3.1% of the nation's gross domestic product) [1]. This staggeringly high monetary cost has likely only escalated over the last 15 years with diminished spending on infrastructure and, furthermore, this number does not include indirect costs associated with the inhibition and control of corrosion, which are likely just as high, if not higher, than the direct costs [1–3].

Fundamentally, corrosion of a metal is caused by a series of electrochemical reactions where concurrent metal dissolution (oxidation) and the reduction of oxygen in the presence of water result in the loss of metal and the re-formation of a more thermodynamically stable metal oxide. The generalized anodic and cathodic reactions can be written as follows:

$$M_{(s)} \rightarrow M^{+}{}_{(aq.)} + e^{-} \tag{6.1}$$

$$O_{2\,(g)} + 2H_2O_{(l)} + 4e^{-} \rightarrow 4OH^{-}{}_{(aq.)} \tag{6.2}$$

eventually yielding a metal oxide upon further reaction. The metal substrate itself serves as the conduit for the electrons; charge compensation between the anodic and cathodic half-cells is further facilitated by ion transport through an external medium. Mitigating corrosion thus fundamentally comes down to inhibiting this sequence of half-cell reactions by either impeding electron or ion transport

Graphene Technology: From Laboratory to Fabrication,
First Edition. Edited by Soroush Nazarpour and Stephen R. Waite.

thereby providing a barrier precluding oxygen and water diffusion, or incorporating a sacrificial layer that reacts preferentially instead of the metal surface. While detailed mechanistic understanding is not yet available, graphene coatings that exhibit promise for corrosion inhibition and are the focus of this contribution are thought to mobilize a combination of these modes.

As a good first approximation, all coatings that are designed to inhibit corrosion of metals can be classified according to the following four mechanisms: (i) cathodic protection; (ii) anodic passivation; (iii) electrolytic inhibition; and (iv) active corrosion inhibition [5–9]. Several particularly effective coatings incorporate multiple modes of action. Figure 6.1 schematically represents the four main modes of corrosion inhibition as well as an additional mode, "self-healing," which is a broader concept that has been differentiated from active corrosion inhibition for the purposes of this discussion. The concept of cathodic protection is based on the idea that deposition of a more electropositive metal will polarize the substrate

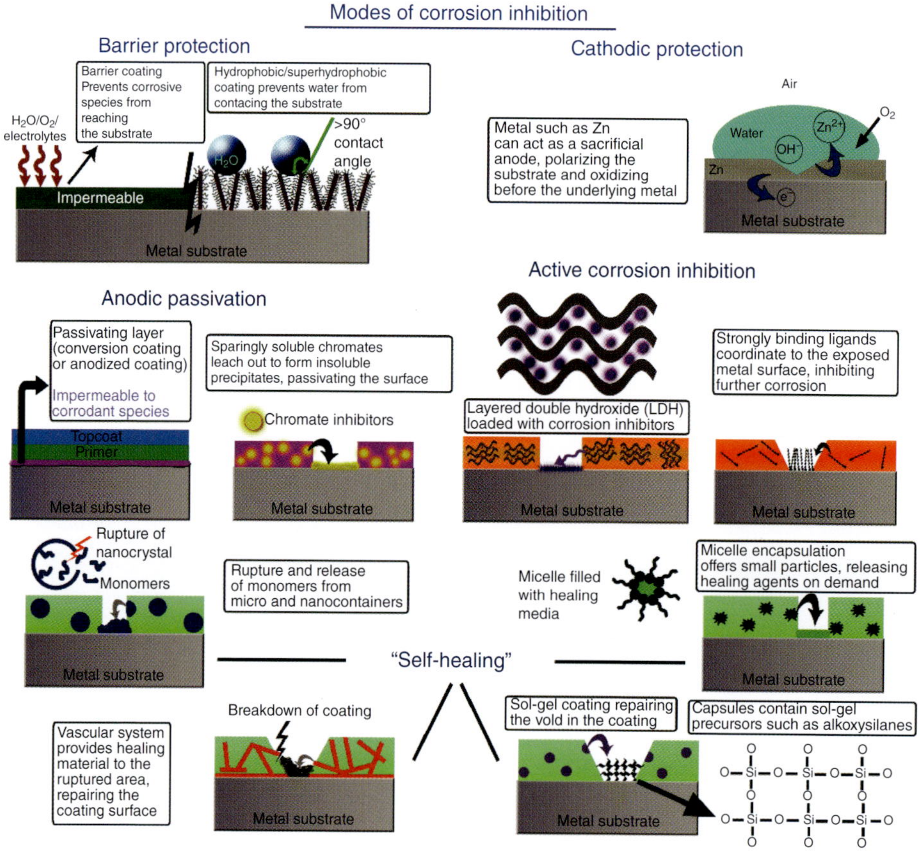

Figure 6.1 Schematic depiction of different modes of corrosion inhibition, including barrier protection, cathodic protection, anodic passivation, active corrosion inhibition, and "self-healing" [9]. (IOP Publishing. Reproduced with permission. All rights reserved.)

metal with the coating thus serving as a sacrificial anode that is preferentially oxidized. The most common example of this mode of action is galvanization of steel or other metal substrates by hot-dip or electroplating processes. Other metals such as aluminum and magnesium have also commonly been alloyed with zinc via interdiffusion or used independently as sacrificial barrier coatings [10]. It is of utmost importance that the sacrificial metal and the substrate be properly coupled so that the polarization of the substrate is sufficient to prevent pitting corrosion while also avoiding "overprotection" that can lead to hydrogen embrittlement or alkaline attack [5].

Anodic passivation of a metal substrate inhibits corrosion by using a passivating layer to coat the metal surface, which has the effect of suppressing the redox reactions listed in Eqs (6.1) and (6.2). This mode essentially involves deposition of a barrier film on the metal surface that is either impervious to ion diffusion or only allows the selective diffusion of specific ions [6, 11]. This approach has been commonly used for a number of decades in the form of chromate conversion coatings and anodized aluminum. In this approach, a natural oxide layer is typically combined with some sort of passivation layer, which together create a bipolar precipitate that strongly retards ion transport through the barrier. Additionally, closely aligned or alternating layers of densely packed oxides and porous oxides deposited during an anodization or conversion coating step redirect ion transport between the anodic and cathodic sites [6, 12, 13].

A third mode of corrosion resistance, electrolytic inhibition, also involves shutting down ion transport pathways between the anodic and cathodic sites of the metal by using a low-ionic-conductivity matrix or diffusion barrier. These coatings are typically barriers that attempt to limit the transport of corrodant electrolyte species to the metal by increasing the tortuosity of the conduction pathways and reducing the movement of charge. Finally, active corrosion inhibition addresses the inevitable scenario of coating failure. Active corrosion inhibition involves the incorporation of components that can be selectively released upon failure or in response to external stimuli, thereby reconstituting a protective barrier at the metal interface [5, 6, 14, 15]. Some examples of these active corrosion inhibitors include strongly binding ligands (with high binding constants for formation of dative bonds at metal surfaces) or sparingly soluble oxide precursors (with low solubility product, K_{sp}, values) that can precipitate, even at very low concentrations, to form barrier layers. Typically, such active species are encapsulated within appropriate polymeric or porous inorganic containers that in turn are embedded within the coating; the active inhibitors are released upon coating failure or the initiation of corrosion [5, 15–18]. The broad umbrella of "self-healing" coatings also includes the incorporation of monomers and catalyst particles that can help to actually reconstruct the existing coating upon failure without necessarily addressing corrosion inhibition. Figure 6.1 schematically depicts the main modes of corrosion inhibition with examples of each.

During the last couple of decades there has been a renewed push toward the development of novel coating systems spurred by several critical imperatives. An ever-aging global infrastructure requires a re-evaluation of current coating technologies with increasing realization that many conventional coating materials represent a hazard to human health and the environment. One of the single most important forces driving the development of new coating technologies is the stringent regulatory environment that addressess the potent carcinogenicity and environmental concerns of hexavalent chromium, which has been used extensively in chromate conversion coatings and hard-chrome electroplating processes for a number of decades. Hexavalent chromium-based coatings have been commonly used in the industry due to their excellent corrosion resistance and self-healing properties (depicted in Figure 6.1); such coatings also provide excellent sheen and metallic luster that gives it its distinctive aesthetic appeal [5, 19–25]. The exceptional mobility of hexavalent chromium ions that make it ideal for corrosion inhibition also allow for the facile environmental transport of these species [19, 22, 26, 27]. Given the increasing realization of the long-term environmental legacy of hexavalent chromium, its use is stringently regulated around the globe [19, 22, 23, 25, 28, 29]. Similarly to hexavalent chromium, volatile organic compounds (VOCs) found in most polymer and paint coatings (as organic solvents and sometimes toxic curing agents) represent a major occupational safety hazard, particularly during application of the coatings [30–32]. Consequently, there is a global push towards sustainable technologies that incorporate earth-abundant and non-toxic components while at the same time providing protection over protracted periods of operation.

Sacrificial coatings such as galvanized zinc have been the "gold standard" for corrosion-resistant coatings for more than a century due to the ease of plating and the excellent barrier and sacrificial coating properties that they bestow upon the underlying base metal substrate [33, 34]. Sacrificial coatings such as zinc owe their corrosion-inhibition properties to the difference in reduction potentials between the sacrificial metal and the substrate metal whereby the coated metal is oxidized preferentially, preventing oxidation of the substrate (as schematically illustrated in Figure 6.1) [33]. Unfortunately, the longevity of the corrosion inhibition is directly proportional to the coating thickness and in a sense these coatings function on the basis of continuous dissolution or failure over their operational lifetime. In other words, ensuring longevity of protection often requires considerable cost and weight penalties with the latter being particularly consequential for aerospace and transportation applications since increased weight oftentimes correlates to increased fuel consumption [9, 35]. Since base metals are for the most part a commodity business with small margins, the recent market unpredictability and the dramatic fluctuations in the price of zinc have been a cause for serious concern in the steel industry and have substantially cut into profit margins [9]. On the other end of the spectrum, high-performance coatings are more critical than ever for advanced lightweight alloys that are increasingly finding use in aerospace and transportation applications. The highly heterogeneous nature of these alloys, oftentimes characterized by intermetallic or elemental precipitates, make them particularly susceptible to corrosion by establishing surface domains with distinctive reduction potentials that can each serve as the cathodic and anodic halves of corrosion cells. To compound this problem, zinc does not have the ability to protect aluminum and aluminum alloys from corrosion since aluminum lies lower in the galvanic series than zinc and thus new, more electroactive, coating systems are urgently required [36].

Given the inadequacies of current technologies, the exploration of entirely new coating paradigms has become of utmost importance. As a carbon-based electroactive and highly conducting material that can be interfaced with metals, graphene is an attractive non-metallurgical candidate for protecting base metals either by itself or as the active element of polymer, metal matrix, or ceramic composites [4, 37]. In this chapter, we review current mechanistic understanding of how graphene inhibits corrosion and provide illustrative examples of the use of graphene as a component in- or as a standalone coating for the corrosion inhibition of base metals. The objective of this contribution is to contrast approaches for corrosion protection based on the use of graphene, explore the mechanistic underpinnings of the protection bestowed by graphene coatings, capture a snapshot of this rapidly developing area of research, and to examine the potential of this material as an alternative to conventional metallurgical films or polymeric coatings. Additionally, the authors' perspective on the outlook for graphene-based corrosion-resistant coatings has been included with a focus on the obstacles to commercialization of graphene: price, consistency, and quality.

6.2 Bare Graphene as a Protective Barrier

6.2.1 Some Electronic Structure Considerations at Graphene/Metal Interfaces

The conduction and valence bands of graphene, derived from p_z orbitals, adopt a conical configuration and intersect at the Fermi level (E_F) [38, 39]. Remarkably, the bands show a linear dispersion as a function of energy ± 1 eV from the point of intersection (the Dirac point). Such a "slim hourglass" electronic structure has some peculiarities that render it particularly useful for protecting metal surfaces. The low density of states near E_F imply that metals (among other species) can readily participate in charge transfer interactions with graphene depending on the relative alignment of their work functions and the extent of overlap of the graphene π-cloud with metal orbitals of the appropriate symmetry [38, 40]. Extensive details of electronic structure consequences of interfacing graphene with metals and dielectrics have been reviewed elsewhere and are summarized here only to provide a perspective of mechanisms for mitigating corrosion [38, 41–43]. As a first approximation, charge transfer between metal surfaces and graphene induces a potential barrier at the graphene/metal interface and the resulting polarization impedes the electron transfer processes necessary for corrosion depicted in Eqs (6.1) and (6.2). Unlike in a bulk solid, charge transfer and other perturbations propagate across the 2D geometric structure and can profoundly alter the electronic structure of graphene [38]. The nature of the metal/graphene interface is thus of paramount importance in determining the extent of polarization and consequently the degree of protection afforded to the metal substrate by graphene.

Several different types of graphene/metal interactions can be distinguished. For low work-function and highly electropositive metals such as Li, Na, K, and Cs, graphene serves as an electron acceptor and a rigid shift of the band structure is observed as a result of electron doping [9, 38]. In contrast, a broad class of transition- and post-transition metals such as Cu, Ag, Au, and Pt exhibit interactions reminiscent of physisorption accompanied by charge transfer and development of an interfacial dipole as illustrated in Figure 6.2. For these metals, *ab initio* density functional calculations predict that while at distant separations, the directionality of charge transfer and the magnitude of the interfacial dipole are well predicted by the relative alignments of work functions, at closer separations, exchange repulsion terms assume greater significance [44, 45]. At a Cu/graphene interface, the direction of charge transfer flips from n- to p-type doping with increasing separation between graphene and the underlying Cu(111) surface [46]. In other words, the repulsion between the itinerant electrons in the π-cloud of graphene and the electron gas of the metal contribute significantly to the surface potential difference (denoted as $\Delta\nu$ in Figure 6.2) that develops at the interface. This potential difference can thus impede redox processes involved in corrosion.

As a third type of metal/graphene interface, for metals such as Ni, Co, Pd, and Ti, the high degree of epitaxial matching of crystal lattices with graphene as well as the strong hybridization of the transition metal d_{z2} orbitals with the graphene π-cloud profoundly reshapes the electronic structure of graphene, opening up a bandgap at the Fermi level and removing spin degeneracy [45–47]. The interfaced graphene layer thus acquires some carbidic character and is rendered a semiconductor [9, 38]. A potential difference now exists between the surficial (and sub-surficial) metal layers that are hybridized with the graphene and constitute a semiconductor and the underlying metallic layers that are relatively unperturbed by interfacing with graphene. The semiconductor/metal interface thus established within the metal gives rise to a Schottky barrier to the tunneling of electrons and this potential barrier can further impede electron transfer at the metal/graphene interface. The height of the barrier depends on the pinning of the Fermi level of the carbide-like semiconductor formed at the graphene/metal interface [48]. Indeed, the Schottky barrier represents a major challenge with making ohmic contacts to semiconducting carbon nanotubes. In other words, both physisorption and covalent hybridization of graphene on the metal surface give rise to interfacial potential barriers that serve to protect against oxidation of the metal.

Another potential mechanism, illustrated in Figure 6.2, derives from the high electrical conductivity of graphene where room-temperature mobilities can readily surpass 10 000 $cm^2 V^{-1} s^{-1}$. The much higher electrical conductivity of graphene as compared to the underlying metal substrate provides an alternative

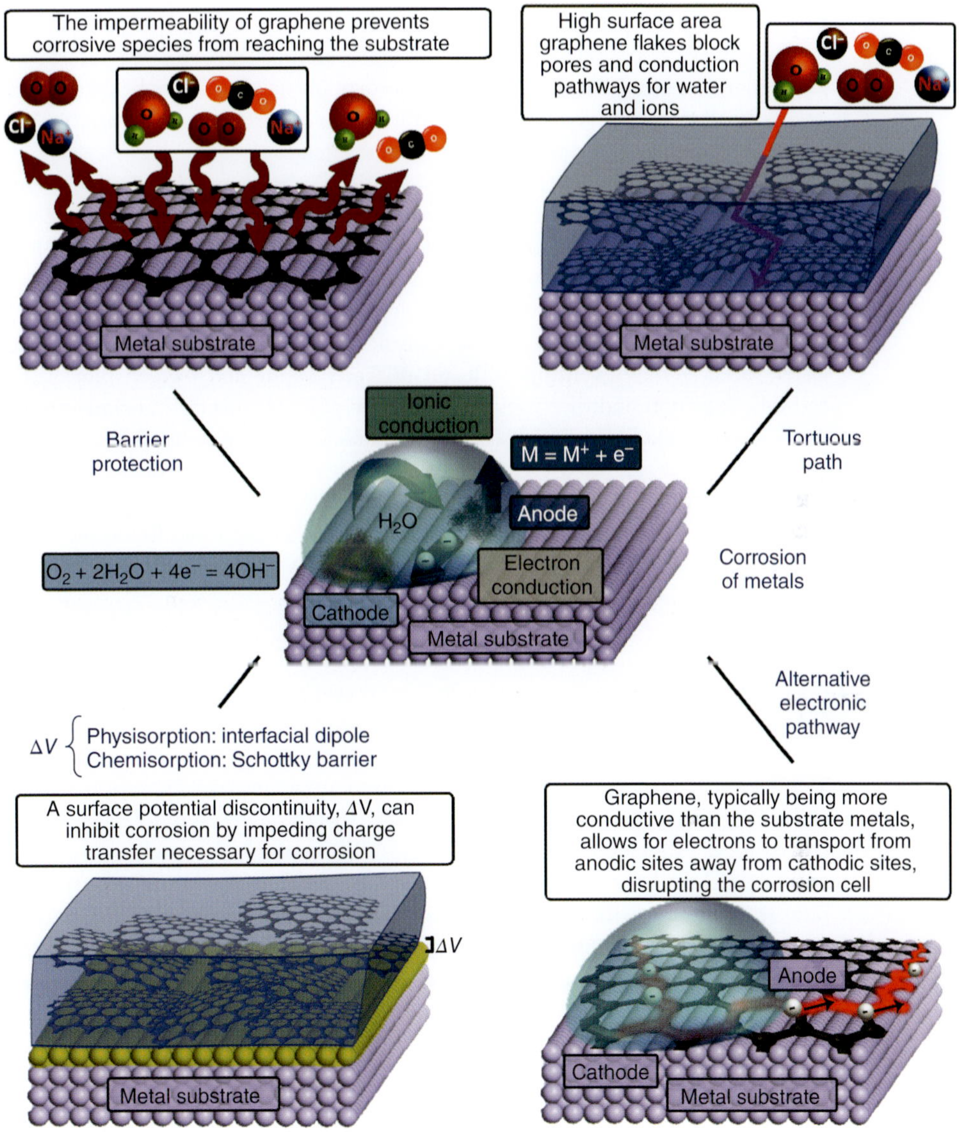

Figure 6.2 Schematic depiction of the four main modes of corrosion inhibition by graphene. This graphic depicts the ways by which graphene can help to impede or entirely shut down the electrochemical processes related to corrosion: by providing barrier protection; by requiring a tortuous path for ion permeation; by formation of a potential barrier at the graphene/metal interface (either a Schottky barrier or interfacial dipole); and by providing an alternative electronic pathway.

conduit for transporting electrons generated at the anode away from cathodic sites thereby disrupting the corrosion reactions [4]. Notably, a mechanism predicated on electrical conductivity is not confined to being an interfacial phenomenon as in the case of direct physisorption or covalent hybridization of graphene on metal surfaces. Indeed, the formation of percolative networks within composite coatings could have much the same effect.

Beyond the electronic structure considerations, it is also primarily the hexagonal honeycomb-like structure of graphene that gives rise to a highly efficient and impervious barrier to diffusion of gaseous and vapor-phase species [49–51]. For instance, Herman and coworkers have shown that CdSe quantum dots with a graphene overlayer show prolonged resistance to photodegradation (and retain their bright luminescence) as a result of limited diffusion of oxygen and water vapor [49]. Preventing diffusion of water vapor and oxygen blocks the cathodic reactions and thus prevents oxidation of metal surfaces. However, even the highest quality graphene layers are characterized by point defects, grain boundaries, and extended cracks, which are inevitable due to their growth on polycrystalline metal foils that themselves have a highly heterogeneous surface structure. The extended defects allow pathways for gaseous and liquid diffusion and thus barrier properties by themselves are not expected to provide protection over prolonged operational lifetimes.

6.2.2 Graphene as a Standalone Corrosion-Resistant Coating and Some Mechanistic Considerations

The degree of protection provided by graphene to a copper substrate has been examined for graphene samples grown directly onto polycrystalline copper foils by chemical vapor deposition (CVD) [52, 53]. Extensive electrochemical studies and Tafel analysis of potentiodynamic plots allow for derivation of corrosion rates from the measured corrosion current density in various corrosive environments, including aqueous solutions of Na_2SO_4 and NaCl [52, 54]. The open circuit potential of graphene-coated Cu is characterized by a slightly lower value (decreased by about 25 mV) and, furthermore, this sample shows a much lower corrosion current density relative to its uncoated counterpart [52]. A corrosion rate of 5.76×10^{-13} m s^{-1} has been deduced for bare copper, and is diminished by a factor of seven to 7.85×10^{-14} m s^{-1} upon coating with graphene when exposed to a 0.1 M Na_2SO_4 electrolyte [52].

Analogous tests have been performed for nickel substrates with and without graphene overlayers [52, 55]. Graphene has been applied onto nickel foils through two distinct methods: (i) direct growth onto the nickel surfaces by CVD from hydrocarbon precursors and (ii) mechanical transfer of graphene grown by CVD onto Cu foils using poly(methyl methacrylate) (PMMA) as the transfer medium [52, 55]. The graphene film grown directly via CVD is observed to provide a higher degree of protection, which is attributed to the high degree of uniformity, relatively pinhole-free coverage, and improved adhesion obtained for these films as compared to mechanically transferred coatings [52, 55]. The latter approach allows for deposition of multilayered films but does not appear to be as effective. The corrosion rate of bare nickel has been determined by Tafel plot analysis to be 2.99×10^{-14} m s^{-1} [52], and is reduced by almost 20-fold to 1.71×10^{-15} m s^{-1} for graphene grown directly on a nickel substrate [52, 55]. Graphene that is mechanically transferred to a nickel substrate provides a substantially reduced degree of protection (only a fourfold reduction); the corrosion rate is 1.83×10^{-14} m s^{-1} for two layers of transferred graphene and 7.62×10^{-15} m s^{-1} for four layers of transferred graphene [52, 55]. Interestingly, the graphene coatings do not substantially alter the negative current potentials corresponding to cathodic reduction of oxygen for either the Cu or Ni samples [52]. This is due in part to graphene being impervious to gas and demonstrates one mechanism by which it can slow the corrosion process [52]. These authors observe that corrosion is initiated in unprotected regions beneath extended defects in graphene. Researchers at Tata Steel have reported that the graphene protection of stainless steel 304 L and 316 L substrates greatly decreases the corrosion rates from 0.7 and 0.5 mm yr^{-1}, respectively, to 1.96×10^{-4} and 8.59×10^{-4} mm yr^{-1}, respectively, in a harsh fuel cell environment [4].

Despite the promising results noted earlier, several researchers have pointed out that due caution is warranted in extrapolating short-term corrosion protection observed in electrochemical tests to prolonged operational lifetimes. In particular, Zottl and coworkers have suggested that even though graphene excels in corrosion tests that capture a snapshot of the ongoing processes (such as cyclic voltammetry and potentiodynamic testing), failure is inevitable over prolonged periods of operation under ambient conditions. Indeed, these researchers suggest that graphene coatings eventually *accelerate* corrosion and that graphene-coated samples fare much worse under ambient conditions as compared to even bare metal substrates [37, 52, 56]. These authors claim that the impermeability of graphene to gas diffusion along with the inevitable defects results in local concentration of corrodant species at the metal interface (through seepage), thereby exacerbating corrosion [56]. In contrast to bare copper, where a native oxide forms uniformly and homogeneously across the surface and affords some degree of protection, upon coating with graphene corrosion is initiated at specific defect sites. The inhomogeneous surface oxide generates stress, resulting in crack formation and generating fresh sites for surface oxidation. The high conductivity of graphene further implies that electrons can be transported across large distances where they can further facilitate cathodic reactions, once again initiating the cycle of surficial and eventually bulk corrosion upon crack formation and propagation [55–57].

It is apparent that at the present time, a single, defect-free graphene layer is difficult to grow, and it is virtually impossible to mechanically transfer while still retaining its uniform and pinhole-free characteristics [38, 52, 56]. Consequently, coatings relying solely on barrier properties of monolayer graphene or few-layered graphene are unlikely to yield corrosion inhibition over prolonged periods of time even under ambient conditions. It is important to note that the challenges here are not *per se* derived from the properties of graphene but instead arise from the inevitable presence of defects, which allows the electrochemistry required for wet corrosion to run rampant, yielding an outcome much worse than for bare copper where the native oxide provides some degree of passivation [56, 57]. These findings suggest that practical coatings must be significantly thicker and must mitigate the accumulation of corrodant species at defect sites. Indeed, coatings developed at Tata Steel as well as in our laboratories either make use of significantly thicker coatings wherein the high surface area of graphene establishes a highly tortuous path for transport of corrodant species or deploy an additional ion impervious matrix [4, 9, 37, 58]. Thicker films of graphene oxide, obtained by oxidation of natural flake graphite with strong oxidizing agents, can be electrophoretically deposited onto metal substrates upon the application of DC fields. The ionized functional groups on graphene oxide provides a surface charge that enables their manipulation and deposition using electric fields [59, 60]. The resulting graphene oxide coating can be reduced (using a suitable reducing agent such as hydrazine or $NaBH_4$) to reduced graphene oxide and initial reports indicate that such coatings provide excellent protection against oxidizing agents [37, 59, 61].

6.3 Graphene Nanocomposites for Corrosion Inhibition

In recent years, polymer-based nanocomposites, wherein an inorganic filler is incorporated within a continuous polymer matrix, have attracted interest for a number of different applications including corrosion-resistant coatings [7, 9, 17, 62]. A particular advantage that these materials offer (subject to overcoming challenges with dispersion and interfacial bonding) is the ability to incorporate modular function wherein the desired properties of both the polymeric matrix and the inorganic filler can be deployed toward some functional purpose. Furthermore, such hybrid composites can often be applied onto surfaces through continuous phase liquid coating or roll-to-roll printing methods, allowing for large-scale industrial deployment [63].

Given the remarkable properties of graphene, both electronic and mechanical, the incorporation of graphene within polymer matrices has emerged as a means to use this material within coatings as an alternative to the direct deposition of graphene onto metal substrates discussed in the preceding section. Furthermore, the need for lightweight and strong composite materials has led to a flurry of research on bulk graphene nanocomposites [64]. Two primary and related challenges with fabricating nanocomposite coating formulations are: (i) the dispersion of graphene within the polymeric matrix and (ii) designing the best possible interface between graphene and the polymer matrix to facilitate not just dispersion but also strong interfacial bonding. A properly engineered interface is imperative to avoid mechanical slippage and aggregation of the graphene within the composite [9]. Graphene has indeed been successfully integrated into a number of polymers over the last decade; a representative (but not exhaustive) list of polymer matrices includes polyaniline (PANI), PMMA, polyetherimide

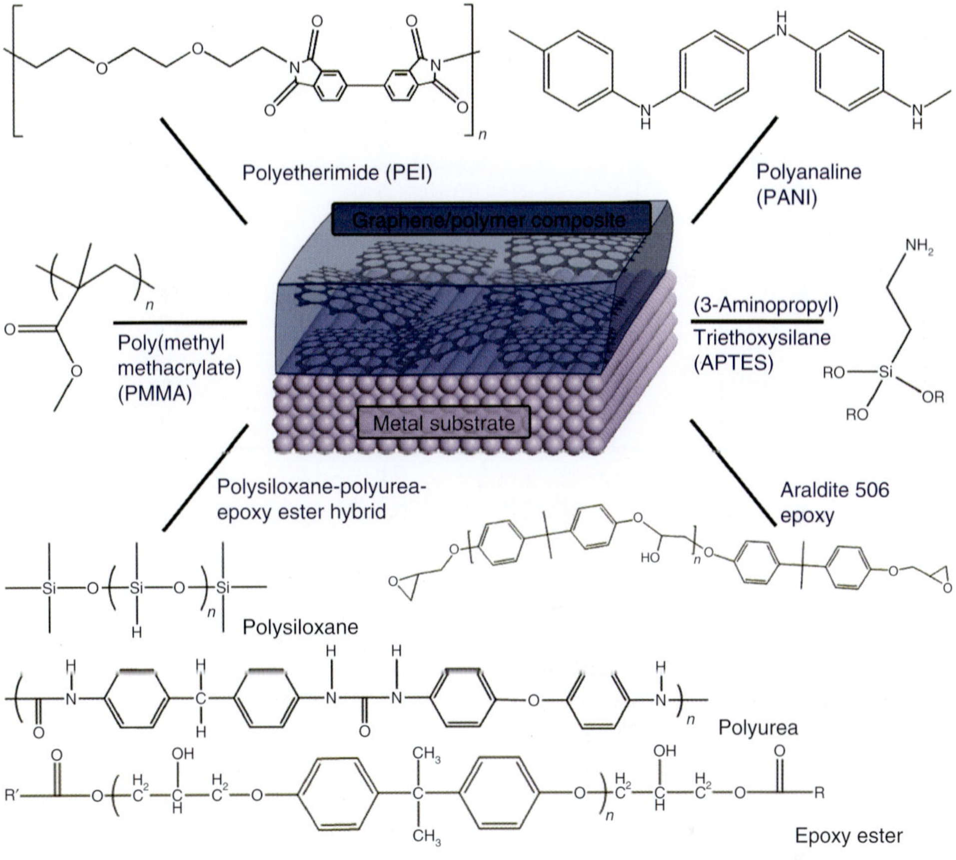

Figure 6.3 Schematic depiction and overview of selected polymers that graphene has been incorporated in for use as a corrosion-resistant coating.

(PEI), various epoxy resins, and polysiloxanes (Figure 6.3) [37, 58, 65–71]. In many cases, these nanocomposites have then been applied to metal substrates for corrosion protection and several orders of magnitude improvements in longevity have been observed. While this specific area of research is relatively new, it holds tremendous promise as a sustainable approach for the inhibition of corrosion [4]. In this section, several types of nanocomposites that show exceptional promise are overviewed.

A notable example of a graphene nanocomposite coating that provides excellent corrosion inhibition to low-alloy steel under salt water exposure was reported previously by the authors of this chapter and involves the incorporation of graphene within a PEI matrix [37]. In this work, graphene is exfoliated from graphite using *N*-methylpyrrolidone (NMP) as the solvent; subsequently, polyamic acid (PAA) is also prepared in NMP and graphene is dispersed in PAA. The final coating is obtained by wire-bar deposition of the graphene/PAA nanocomposite onto a freshly degreased low-alloy steel surface followed by imidization at 250 °C to

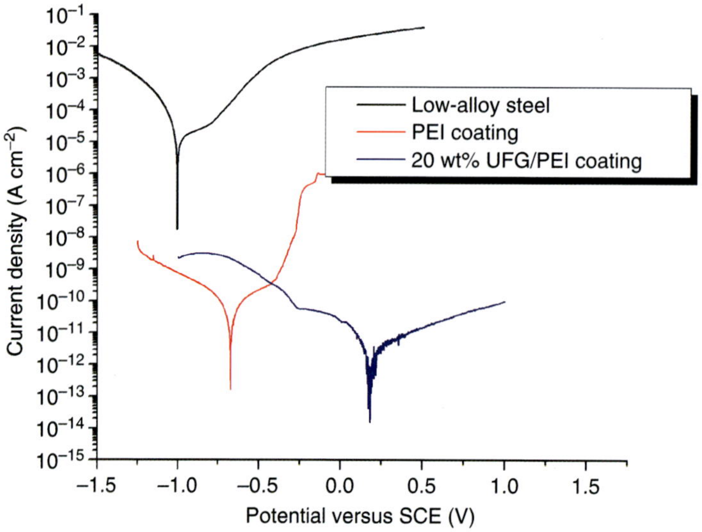

Figure 6.4 Tafel plot showing enhanced corrosion resistance afforded by the 20 wt% UFG/PEI coating as compared to a PEI coating and uncoated low-alloy steel.

obtain PEI. A 20 wt% graphene (UFG)/PEI coating showed a significant drop in corrosion current density of approximately seven orders of magnitude from the blank steel as well as the formation of a passivation band in the Tafel analysis of potentiodynamic polarization data (Figure 6.4). Based on the data presented in Figure 6.4 and summarized in Table 6.1, the estimated corrosion rate for a low alloy steel is decreased by six orders of magnitude as compared to the blank low-alloy steel and about an order of magnitude as compared to the PEI coating alone [37]. Extended exposure tests (more than 3100 h in 3.5% NaCl) of this coating system on cold-rolled steel showed qualitatively that the lifetime of the graphene coating was significantly greater than that of the PEI alone (Figure 6.5). As a result of the excellent dispersion of the graphene within the PEI, likely owing to favorable $\pi-\pi$ stacking interactions, a percolative network of graphene appears to be established within the coating that provides an alternative electronic pathway to prevent electrons from flowing from the anodic to cathodic half-cells. In addition, the well-dispersed high-surface-area graphene platelets impose a

Table 6.1 Corrosion potential (E_{corr}), corrosion current density (I_{corr}), and extrapolated corrosion rate in mm yr^{-1} from electrochemical testing of low-alloy steel, PEI coating, and 20 wt% UFG/PEI coating.

Sample	E_{corr} (V)	I_{corr} (A cm^{-2})	Corrosion rate (mm yr^{-1})
Low-alloy steel	−1.01	1.29×10^{-5}	9.53×10^{-3}
PEI coating	−0.67	7.00×10^{-11}	5.19×10^{-8}
20 wt% UFG/PEI coating	0.19	3.00×10^{-12}	2.22×10^{-9}

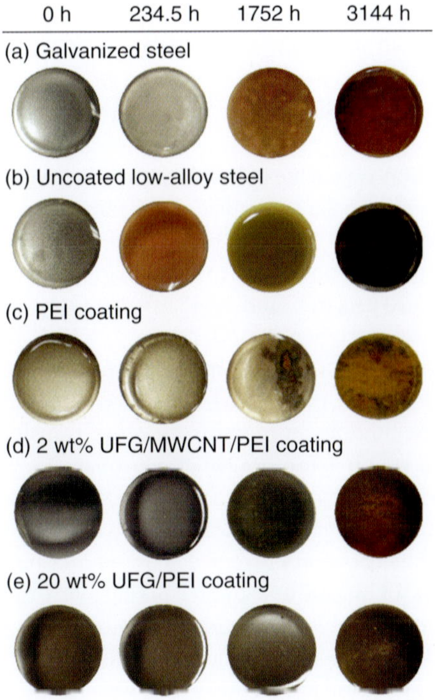

Figure 6.5 Digital photographs of salt-water immersion measurements on uncoated low-alloy steel, a PEI coating, and a 20 wt% UFG/PEI coating [37]. *American Ceramic Society Bulletin*. Used with permission.

tortuous path for ion permeation and further likely yield a potential barrier at the metal interface (Figure 6.2) [37].

In an alternative approach developed by the authors of this work, graphene oxide was used both as the active filler material and as the curing agent and covalently linked to the host polymer matrix, thereby achieving excellent dispersion at high loading levels [68]. A commercially available epoxy resin, Araldite 506, was used as the matrix. This resin typically requires a curing agent and/or elevated temperatures to initiate the reaction; however, the hydroxyl and carboxylic acid groups of graphene oxide can react with the epoxide groups of the Araldite, thereby constituting a cross-linked network. This approach enabled the incorporation of graphene loadings up to 50 wt% within the epoxy matrix. Samples with high loading levels of graphene oxide exhibited excellent corrosion resistance upon exposure to 3.5 wt% aqueous solutions of NaCl even after 2160 h of exposure. Lower loading levels (e.g., 5 wt%) did not perform as well in the extended exposure tests, likely as a result of the lower cross-linking density (which renders the matrix more permeable to water and corrodant species) as well as the relatively low amount of the electro-active filler material [68].

Several analogous approaches have been developed to improve the interfacial chemistry between graphene and the polymer matrix including that of Chang *et al.* demonstrated that functionalizing graphene sheets with 4-aminobenzoic acid allows for their facile incorporation within PANI [65]. PANI by itself is one of most promising polymeric systems used for corrosion protection of steels as a result of its high electroactivity and facile charge transfer with steel substrates [72]. Unfortunately, it can be difficult to achieve good dispersion of graphene within PANI; however, excellent dispersion was achieved with the mediation of 4-aminobenzoic acid and authors report excellent corrosion-resistant barrier properties. Tafel analysis for this system showed a significant drop in corrosion current density and a shift to a more positive potential from the bare steel or that of the PANI coating alone (from 3.70 $\mu A\ cm^{-2}$ and −647 mV for PANI to 0.38 $\mu A\ cm^{-2}$ and −537 mV for a 0.5 wt% graphene/PANI coating), which in turn also decreased the estimated corrosion rate by approximately two orders of magnitude [65]. The incorporation of graphene within PMMA has been achieved using a "grafting from" approach by first binding an initiator for atom-transfer radical polymerization, (*N*-(2-aminoethyl)-2-bromo-2-methylpropanamide, NABM), and subsequently polymerizing methyl methacrylate (MMA) from the surface of graphene [69]. Graphene oxide loadings of up to 81 wt% within PMMA are achieved by this method; the corrosion current density of a coated copper sheet decreased by 3–4 orders of magnitude upon coating with the graphene/PMMA composite. The coating was able to provide corrosion protection to the copper surface even up to 100 h of exposure to the 3.5% NaCl solution [69]. As a concluding example, Okafor *et al.* found that graphene that was incorporated within a hybrid polymer of epoxy ester, siloxane, and urea afforded excellent corrosion protection properties for Al coupons [67]. Even at concentrations as low as 1–2 wt% of graphene it was possible to achieve a significant diminution in corrosion, as corroborated by electrochemical impedance spectroscopy (EIS) and Tafel analysis of potentiodynamic polarization experiments. Tafel analysis of polarization data indicates a drop in the corrosion current density of about two orders of magnitude as compared to the neat hybrid polymer coating [67]. From a practical perspective, graphene/polymer nanocomposites likely represent the most viable option for large-scale applications given the drawbacks of single- and few-layered graphene as stand-alone coatings summarized in the preceding section. The widespread use of nanocomposites incorporating carbon black, microstructured carbon, carbon fibers, and, increasingly, carbon nanotubes, renders the incorporation of graphene within polymers a relatively facile "drop-in" solution for a diverse range of paints and coatings. The static dissipation properties arising from the formation of a percolative network are an added bonus, particularly for packaging applications in the semiconductor industry.

6.4 Graphene/Metal Nanocomposites for Corrosion Inhibition

The incorporation of graphene within metallurgical thin films yields interesting metal matrix composites some of which hold promise for corrosion protection [73]. Metallurgical films and alloys are extremely effective at preventing corrosion of various metal substrates and can serve as sacrificial anodes when used as coatings. Alternatively, metals that lie higher in the reductive potential series can also serve to protect substrates by creating a barrier that requires an increased potential to initiate oxidation. Incorporation of graphene (or carbon nanotubes) can increase the formability of the metal coating and potentially offer enhanced electrical conductivity. The primary drawback is that dispersion of graphene within metals remains a formidable challenge and the dissimilar graphene/metal interface is prone to debonding, which in turn can lead to incipient porosity and serve as the nucleation point for initiation of corrosion [73]. This drawback limits the amount of graphene that can be incorporated within such composites.

Corrosion studies have been performed on several graphene/metal composite materials including Ni/graphene [74, 75], Zn/graphene [76], and Sn/graphene [77]. These coatings were all synthesized via combined electrophoretic deposition and electroplating from dispersions of chemically modified graphene or exfoliated graphene in plating baths. The substrate used in all cases was mild steel. Based on analysis of X-ray diffraction data, the incorporation of graphene brings about a substantial change in the texture and grain size of the electrodeposited films. Table 6.2 collates the percentage decrease in grain size reported by various authors [74–77]. While the values for each trial cannot be compared directly, the general trends suggest a pronounced decrease of grain size upon incorporation of graphene. The decrease in grain size can be attributed to an increased density of nucleation sites on the metal surface as a result of the presence of graphene, which could potentially inhibit grain growth. Changes in texture could also potentially result from changes in the preferred crystallographic growth planes during deposition as a result of deposition onto graphene and not metal surfaces. Kumar and Berlia have postulated that graphene could limit access of solution phase metal to the substrate surface, thereby limiting particle growth [76, 77]. Several of these studies also indicate that the surfaces of the coatings exhibit protrusions of hillock structures upon the incorporation of graphene. These protrusions

Table 6.2 Average grain size calculated using the Scherrer equation and the reduction in grain size upon incorporation of graphene [74–77].

	Grain size (nm)		Reduction in grain size (%)
	Metal	Composite	
Ni [74]	30	20	33
Ni [75]	35	~19	46
Ni [75]	35	~16	54
Zn [76]	70	62	11
Sn [77]	79.46	75	5.6

Table 6.3 Summary of Tafel analysis results for Ni/graphene, Zn/graphene, and Sn/graphene composite coatings [74–77].

	I_{corr} (μA cm^{-2})		E_{corr} (V)	
	Metal	Composite	Metal	Composite
Ni [74]	15.8	6.687	−0.492	−0.398
Ni [75]	19.1	3.02	−0.2665	−0.2512
Ni [75]	19.1	0.398	−0.2665	−0.2346
Zn [76]	19.86	6.82	0.915	0.920
Sn [77]	1.365	0.815	−0.573	−0.537

suggest the partial segregation of graphene domains and can potentially be sites for initiation of failure.

The metal matrix composites show an appreciable increase of microhardness, which can be attributed to the high strength of graphene as well as the small grain size achieved in the electrodeposited films. The diminution in grain size prevents the build-up and movement of dislocations [75]. Kumar *et al.* suggest that despite the presence of hillocks, the overall density of pits is reduced in Zn/graphene coatings upon graphene incorporation, likely as a result of the ability of graphene to bridge or fill gaps [76]. The reduced density of pits is thought to be favorable for the protection of the underlying metal substrates since such defects can serve to initiate corrosion. In each study, Tafel analysis was performed on electrodeposited bare metal and the metal/graphene composite films. The results of these studies are summarized in Table 6.3. In all cases, the corrosion current density (I_{corr}) was observed to substantially decrease for the composite material and the corrosion potential (E_{corr}) became more positive. The corrosion rate for the Zn/graphene composite decreased fourfold as compared to the bare metal; similarly, the corrosion rate for the Sn/graphene composite was about 60% of the value for the bare Sn film.

Graphene oxide has also been used instead of reduced graphene oxide within cobalt/graphene oxide composite coatings deposited by electrodeposition onto mild steel [78]. Graphene oxide can be used with a wider variety of solvents when compared to reduced graphene oxide and the oxygen groups present on the surface allow for a greater range of coordinative interactions with a metal substrate. Analogous to the results summarized in Table 6.2, the grain size of cobalt was reduced to 20 ± 2 nm for a cobalt/graphene oxide composite coating compared to a value of 50 ± 5 nm for the bare cobalt film [78]. These authors also observed a change in the preferred growth orientation of the cobalt films upon incorporation of graphene. Tafel plot analysis indicated corrosion potentials of −0.3149 and −0.3597 V for the bare cobalt and cobalt/graphene oxide composite coatings, respectively. The corrosion current decreased from 9.70×10^{-6} A cm^{-2} for the bare cobalt films to 3.04×10^{-6} A cm^{-2} for the cobalt/graphene oxide coatings. The corrosion rate also decreased similarly for the composite, going from 4.98×10^{-2} mm yr^{-1} for the bare metal coating to 1.56×10^{-2} mm yr^{-1} for the cobalt/graphene oxide coating [78].

Graphene/metal nanocomposites thus show promising performance for corrosion inhibition. Graphene is observed to decrease the grain size of electroplated metal films and to increase their hardness. By providing more nucleation sites and enabling homogeneous deposition with reduced pit density, the nanocomposite coatings increase the resistance of the electrodeposited films to corrosive attack. Further optimization of graphene/metal interfaces is clearly required to mitigate a major probable reason for failure.

6.5 Graphene/Ceramic Nanocomposites for Corrosion Inhibition

Several reports in the literature suggest the attachment of silica or alumina prior to incorporation within an epoxy resin to not only disperse the material more readily, but also in some cases to prevent the graphene from having any deleterious effects on the corrosion resistance [58, 66, 70, 71]. Some research has suggested that because graphene is so electro-active it could actually increase the degree of corrosion for the underlying metal by increasing the number of active cathode sites throughout the coating matrix, essentially setting up a graphene/metal couple that causes galvanic corrosion [56, 70, 79]. Consequently, a number of different approaches have been developed to incorporate graphene within a number of commercially available epoxy resins wherein the graphene is first encapsulated by an insulating ceramic layer [66, 70, 71, 79, 80]. The enhancement in corrosion resistance that is observed for such composites is predicated entirely on the high surface area and aspect ratio of the graphene, which implies that any corrosive species must take a tortuous path to reach the metal surface (Figure 6.2). As a representative example of this approach, Sun *et al.* showed that graphene encapsulated by a nanometer-sized layer of SiO_2 created an effective barrier-type coating when dispersed within a polymer matrix, thereby significantly increasing the corrosion protection afforded to the underlying copper substrates [70].

Alternatively, Aneja *et al.* recently described an approach where they used silica not as an electrically insulating barrier but instead as a means to attach the graphene to a steel surface [58]. In this work, the authors used (3-aminopropyl)triethoxysilane (APTES) to create a functionalized graphene surface that could then be reacted with the steel surface to create a graphene/silica composite that was bonded to hydroxyl groups on the steel substrate. The subsequent deposition of an epoxy layer further enhances the corrosion-resistant properties through its barrier characteristics. Remarkably, potentiodynamic polarization measurements indicate that these silica-functionalized graphene coatings are superior even to samples that have undergone conventional pretreatment with chrome. The corrosion current density of the silica/graphene coating was approximately three orders of magnitude lower than that of bare steel alone and an order of magnitude lower than the chromium-pretreated sample. In addition, a pronounced shift to more positive corrosion potentials was observed for the silica/graphene/epoxy composites. Salt fog testing of these samples as well as EIS data corroborate the results of Tafel analysis, confirming the formation of an excellent corrosion-resistant coating [58].

Another study by Khalil *et al.* examined the corrosion inhibition afforded by a nickel/graphene/anatase-TiO_2 coating on mild steel [81]. This coating design seeks to combine the desirable properties of graphene/metal composites discussed in the preceding section with the possibility for a further increase of stability and increase in hardness through the incorporation of ceramic materials. Graphene/TiO_2 composites were prepared separately prior to being electrodeposited from a plating bath onto mild steel alongside metallic nickel. A coating thickness of 20 μm was achieved with a concentration of $0.4\,g\,L^{-1}$ graphene-TiO_2 particles in the plating bath at a current density of $10\,mA\,cm^{-2}$. X-ray diffraction analysis indicates an average particle size of 24 nm for a bare nickel coating and 20 nm for the nickel/graphene/TiO_2 composite deposited under the same conditions. No major changes in the preferred orientation of growth were observed in the diffraction data. Tafel analysis showed a decrease in corrosion current density and corrosion rate with an increase in coating thickness. When comparing films of the maximum thickness, the corrosion current density decreased from $1.53 \times 10^{-6}\,A\,cm^{-2}$ for the nickel/graphene composite to $3.46 \times 10^{-8}\,A\,cm^{-2}$ for the nickel/graphene/TiO_2 composite. The corrosion rate also decreased from $0.0183\,mm\,yr^{-1}$ for nickel/graphene to $0.0004\,mm\,yr^{-1}$ for the nickel/graphene/TiO_2 coating. Nyquist plots showed an increase in polarization resistance from $3.90 \times 10^3\,\Omega$ for nickel/graphene on mild steel to $3.31 \times 10^4\,\Omega$ for nickel/graphene/TiO_2.

The inclusion of ceramic components can thus enhance the corrosion resistance of the first two classes of coatings discussed in this chapter. For graphene/polymer composites, ceramic components can facilitate the immobilization of graphene to SiO_2 surfaces or mitigate graphene/steel galvanic couples. For graphene/metal composites, the inclusion of ceramic components can further increase the hardness of the coatings and tortuosity of the ion permeation pathways.

6.6
Summary and Future Outlook

In this chapter, we have attempted to capture a snapshot of a rapidly emerging discipline that could potentially yield some of the first large-scale commercial products incorporating graphene. Protecting base metals is an urgent imperative not just to prolong the longevity of infrastructure but also to facilitate the use of base metals in many emerging applications related to clean energy. Concerns regarding ecological toxicity of conventional coating materials have created interesting opportunities for the adoption of new technologies. The interesting results obtained for graphene composites suggest that this material could be ideally poised for widespread industrial deployment. It is clear that the use of graphene by itself is unlikely to be practical over prolonged periods of operation given the inevitable presence of extended defects and, therefore, the rational design of polymer, ceramic, and metal composites remains a critical imperative.

Mechanistic understanding of how graphene reacts with metal substrates remains incomplete, particularly with regard to the influence of additional polymeric, metal, or ceramic matrices. In this work, we have discussed several mechanistic possibilities including the development of interfacial potential barriers as a result of exchange repulsions or covalent hybridization, imposition of a highly tortuous path for ion permeation as a result of its high surface area, disruption of electron transport from the anode to the cathode as a result of its high electrical conductivity, and the establishment of an impermeable barrier as a result of its tightly packed covalently bonded structure (Figure 6.2). It is noteworthy that several graphene/polymer and graphene/metal matrix composites show excellent corrosion inhibition not just under potentiodynamic and EIS testing but also over prolonged exposure to accelerated testing environments.

A major challenge in the discipline is the wide diversity of materials that are designated as being graphene, spanning a broad range of thicknesses, lateral dimensions, and extent of functionalization. The inconsistencies noted in the literature possibly derive in large measure from the widely heterogeneous materials used within coatings. With increased quality and consistency of available graphene materials, it is expected that more definitive answers will become available regarding the modes of action. The availability of higher quality and consistent graphene materials will also facilitate the development of systematic structure–function correlations (including through high-throughput experimental methods coupled with the appropriate data analytics), which will subsequently inform optimization of coating formulations much in the same way that integrated computational materials engineering approaches are currently being used for the development of new alloys.

A second major problem pertains to the mode of application. Most studies thus far use laboratory equipment and testing under controlled conditions. For deployment on an industrial scale, methods such as wet casting, electrodeposition, and roll-to-roll printing will need to be developed (which further requires the availability of large amounts of high-quality samples). The results of field testing are yet to become publicly available although it is clear that several companies have now advanced graphene coatings to pilot-plant and field-testing scales. As with other areas of coatings research, the development of multifunctional and "self-healing" coatings is a particularly attractive frontier. The authors hope that this contribution will provide further impetus to this nascent discipline that shows exceptional promise for rapid commercialization.

Acknowledgments

We gratefully acknowledge Dr Tapan K. Rout and Dr Anil Gaikwad of Tata Steel for helpful discussions. Our work on graphene nanocomposite coatings was funded by the New York State Pollution Prevention Institute, Tata Steel, and the Indo-US Science and Technology Forum.

References

1. Koch, G.H., Brongers, M.P.H., Thompson, N.G., Virmani, Y.P., and Payer, J.H. (2002) *Corrosion Costs and Preventive Strategies in the United States*, FHWA, Washington, D.C.
2. National Research Council (2011) *Research Opportunities in Corrosion Science and Engineering*, The National Academies Press, Washington, DC.
3. Winkleman, A., Svedberg, E.B., Schafrik, R.E., and Duquette, D.J. (2011) *Adv. Mater. Process.*, **169**, 26–31.
4. Böhm, S. (2014) *Nat. Nanotechnol.*, **9**, 741–742.
5. Presuel-Moreno, F., Jakab, M.A., Tailleart, N., Goldman, M., and Scully, J.R. (2008) *Mater. Today*, **11**, 14–23.
6. Hughes, A.E., Cole, I.S., Muster, T.H., and Varley, R.J. (2010) *NPG Asia Mater.*, **2**, 143–151.
7. Figueira, R.B., Silva, C.J.R., and Pereira, E.V. (2014) *J. Coatings Technol. Res.*, **12**, 1–35
8. Zheludkevich, M.L., Salvado, I.M., and Ferreira, M.G.S. (2005) *J. Mater. Chem.*, **15**, 5099–5111.
9. Dennis, R.V., Patil, V., Andrews, J.L., Aldinger, J.P., Yadav, G.D., and Banerjee, S. (2015) *Mater. Res. Express*, **2**, 032001.
10. Prosek, T., Nazarov, A., Bexell, U., Thierry, D., and Serak, J. (2008) *Corros. Sci.*, **50**, 2216–2231.
11. Worley, B.C., Ricks, W.A., Prendergast, M.P., Gregory, B.W., Collins, R., Cassimus, J.J., and Thompson, R.G. (2013) *Langmuir*, **29**, 12969–12981.
12. Cole, I.S., Muster, T.H., Lau, D., Wright, N., and Azmat, N.S. (2010) *J. Electrochem. Soc.*, **157**, C213–C222.
13. Sato, N. (1989) *Corrosion*, **45**, 354–368.
14. Clark, W.J., Ramsey, J.D., McCreery, R.L., and Frankel, G.S. (2002) *J. Electrochem. Soc.*, **149**, B179–B185.
15. Grigoriev, D., Akcakayiran, D., Schenderlein, M., and Shchukin, D. (2014) *Corrosion*, **70**, 446–463.
16. Shchukin, D.G., Zheludkevich, M., and Möhwald, H. (2006) *J. Mater. Chem.*, **16**, 4561.
17. Lamaka, S.V., Zheludkevich, M.L., Yasakau, K.A., Serra, R., Poznyak, S.K., and Ferreira, M.G.S. (2007) *Prog. Org. Coat.*, **58**, 127–135.
18. Zheludkevich, M.L., Shchukin, D.G., Yasakau, K.A., Möhwald, H., and Ferreira, M.G.S. (2007) *Chem. Mater.*, **19**, 402–411.
19. Kendig, M.W. and Buchheit, R.G. (2003) *Corrosion*, **59**, 379–400.
20. Groshart, E. (1997) *Met. Finish.*, **95**, 70–72.
21. Bibber, J.W. (2007) *J. Appl. Surf. Finish.*, **2**, 273–275.
22. Bibber, J.W. (2008) *Met. Finish.*, **106**, 41–46.
23. Bibber, J.W. (2011) in *Management of Hazardous Residues Containing Cr (VI)* (ed M.J. Balart Murria), Nova Science Publishers, Inc., Hauppauge, NY, USA, pp. 37–49.
24. Chen, T., Li, W., and Cai, J. (2011) *RSC Adv.*, **1**, 607–610.
25. Osborne, J.H. (2001) *Prog. Org. Coat.*, **41**, 280–286.
26. Shchukin, D.G., Zheludkevich, M., Yasakau, K., Lamaka, S., Ferreira, M.G.S., and Möhwald, H. (2006) *Adv. Mater.*, **18**, 1672–1678.
27. Katz, S.A. and Salem, H. (1993) *J. Appl. Toxicol.*, **13**, 217–224.
28. Anon (1998 *Edited by Agency for Toxic Substances and Disease Registry,*) , Department of Health and Human Services, Atlanta, GA.
29. Baral, A. and Engelken, R.D. (2002) *Environ. Sci. Policy*, **5**, 121–133.
30. Jagtap, S.D., Tambe, S.P., Choudhari, R.N., and Mallik, B.P. (2014) *Prog. Org. Coat.*, **77**, 395–402.
31. Kou, Y., Wang, J., Lv, M., and Jian, X. (2010) *Polym. Int.*, **59**, 107–111.
32. Elmore, J.D., Kincaid, D.S., Komar, P.C., and Nielsen, J.E. (2002) *J. Coating. Technol.*, **74**, 63–72.

33. Marder, A.R. (2000) *Prog. Mater Sci.*, **45**, 191–271.
34. Hosking, N.C., Ström, M.A., Shipway, P.H., and Rudd, C.D. (2007) *Corros. Sci.*, **49**, 3669–3695.
35. Rout, T.K., Gaikwad, A.V. and Dingemans, T. (2011) *Patent Specification* 2011035920A1
36. Dennis, R.V., Viyannalage, L.T., Aldinger, J.P., Rout, T.K., and Banerjee, S. (2014) *Ind. Eng. Chem. Res.*, **53**, 18873–18883.
37. Dennis, R.V., Viyannalage, L.T., Gaikwad, A.V., Rout, T.K., and Banerjee, S. (2013) *Am. Ceram. Soc. Bull.*, **92**, 18–24.
38. Schultz, B.J., Dennis, R.V., Lee, V., and Banerjee, S. (2014) *Nanoscale*, **6**, 3444–3466.
39. Novoselov, K.S., Geim, A.K., Morozov, S.V., Jiang, D., Katsnelson, M.I., Grigorieva, I.V., Dubonos, S.V., and Firsov, A.A. (2005) *Nature*, **438**, 197–200.
40. Sarkar, S., Moser, M.L., Tian, X., Zhang, X., Al-Hadeethi, Y.F., and Haddon, R.C. (2014) *Chem. Mater.*, **26**, 184–195.
41. Batzill, M. (2012) *Surf. Sci. Rep.*, **67**, 83–115.
42. Bekyarova, E., Sarkar, S., Wang, F., Itkis, M.E., Kalinina, I., Tian, X., and Haddon, R.C. (2013) *Accounts Chem. Res.*, **46**, 65–76.
43. Dahal, A. and Batzill, M. (2014) *Nanoscale*, **6**, 2548–2562.
44. Lee, V., Park, C., Jaye, C., Fischer, D.A., Yu, Q., Wu, W., Liu, Z., Bao, J., Pei, S.-S., Smith, C., Lysaght, P., and Banerjee, S. (2010) *J. Phys. Chem. Lett.*, **1**, 1247–1253.
45. Giovannetti, G., Khomyakov, P.A., Brocks, G., Karpan, V.M., van den Brink, J., and Kelly, P.J. (2008) *Phys. Rev. Lett.*, **101**, 026803.
46. Khomyakov, P.A., Giovannetti, G., Rusu, P.C., Brocks, G., van den Brink, J., and Kelly, P.J. (2009) *Phys. Rev. B*, **79**, 195425.
47. Abtew, T., Shih, B.-C., Banerjee, S., and Zhang, P. (2013) *Nanoscale*, **5**, 1902–1909.
48. Tersoff, J. (1984) *Phys. Rev. Lett.*, **52**, 465–468.
49. Zhang, D., Wang, D.Z.-R., Creswell, R., Lu, C., Liou, J., and Herman, I.P. (2015) *Chem. Mater.*, **27**, 5032–5039.
50. Leenaerts, O., Partoens, B., and Peeters, F.M. (2008) *Appl. Phys. Lett.*, **93**, 193107.
51. Bunch, J.S., Verbridge, S.S., Alden, J.S., van der Zande, A.M., Parpia, J.M., Craighead, H.G., and McEuen, P.L. (2008) *Nano Lett.*, **8**, 2458–2462.
52. Prasai, D., Tuberquia, J.C., Harl, R.R., Jennings, G.K., and Bolotin, K.I. (2012) *ACS Nano*, **6**, 1102–1108.
53. Singh Raman, R.K., Chakraborty Banerjee, P., Lobo, D.E., Gullapalli, H., Sumandasa, M., Kumar, A., Choudhary, L., Tkacz, R., Ajayan, P.M., and Majumder, M. (2012) *Carbon*, **50**, 4040–4045.
54. Kirkland, N.T., Schiller, T., Medhekar, N., and Birbilis, N. (2012) *Corros. Sci.*, **56**, 1–4.
55. Singh Raman, R.K. and Tiwari, A. (2014) *JOM*, **66**, 637–642.
56. Schriver, M., Regan, W., Gannett, W.J., Zaniewski, A.M., Crommie, M.F., and Zettl, A. (2013) *ACS Nano*, **7**, 5763–5768.
57. Keil, P., Frahm, R., and Lützenkirchen-Hecht, D. (2010) *Corros. Sci.*, **52**, 1305–1316.
58. Aneja, K.S., Bohm, S., Khanna, A., and Bohm, M. (2015) *Nanoscale*, **7**, 17879–17888.
59. Sahu, S.C., Samantara, A.K., Seth, M., Parwaiz, S., Singh, B.P., Rath, P.C., and Jena, B.K. (2013) *Electrochem. Commun.*, **32**, 22–26.
60. Lee, V., Whittaker, L., Jaye, C., Baroudi, K.M., Fischer, D.A., and Banerjee, S. (2009) *Chem. Mater.*, **21**, 3905–3916.
61. Lih, E.T.Y., Zaid, R.b.M., Ling, T.L., and Chong, K.F. (2012) *Int. J. Chem. Eng. Appl.*, **3**, 453.
62. Bierwagen, G., Brown, R., Battocchi, D., and Hayes, S. (2010) *Prog. Org. Coat.*, **67**, 195–208.
63. Orloff, N.D., Long, C.J., Obrzut, J., Maillaud, L., Mirri, F., Kole, T.P., McMichael, R.D., Pasquali, M., Stranick, S.J., and Alexander Liddle, J. (2015) *Sci. Rep.*, **5**, 17019.
64. Kim, H., Abdala, A.A., and Macosko, C.W. (2010) *Macromolecules*, **43**, 6515–6530.
65. Chang, C.-H., Huang, T.-C., Peng, C.-W., Yeh, T.-C., Lu, H.-I., Hung, W.-I., Weng,

C.-J., Yang, T.-I., and Yeh, J.-M. (2012) *Carbon*, **50**, 5044–5051.

66. Ma, Y., Di, H., Yu, Z., Liang, L., Lv, L., Pan, Y., Zhang, Y., and Yin, D. (2015) *Appl. Surf. Sci.*, **360**, 936–945
67. Okafor, P.A., Singh-Beemat, J., and Iroh, J.O. (2015) *Prog. Org. Coat.*, **88**, 237–244.
68. Patil, V., Dennis, R.V., Rout, T.K., Banerjee, S., and Yadav, G.D. (2014) *RSC Adv.*, **4**, 49264–49272.
69. Qi, K., Sun, Y., Duan, H., and Guo, X. (2015) *Corros. Sci.*, **98**, 500–506.
70. Sun, W., Wang, L., Wu, T., Pan, Y., and Liu, G. (2015) *J. Mater. Chem. A*, **3**, 16843–16848.
71. Yu, Z., Di, H., Ma, Y., Lv, L., Pan, Y., Zhang, C., and He, Y. (2015) *Appl. Surf. Sci.*, **351**, 986–996.
72. Rout, T.K., Jha, G., Singh, A.K., Bandyopadhyay, N., and Mohanty, O.N. (2003) *Surf. Coat. Tech.*, **167**, 16–24.
73. Chen, L.-Y., Konishi, H., Fehrenbacher, A., Ma, C., Xu, J.-Q., Choi, H., Xu, H.-F., Pfefferkorn, F.E., and Li, X.-C. (2012) *Scripta Mater.*, **67**, 29–32.
74. Kumar, C.M.P., Venkatesha, T.V., and Shabadi, R. (2013) *Mater. Res. Bull.*, **48**, 1477–1483.
75. Jiang, K., Li, J., and Liu, J. (2014) *RSC Adv.*, **4**, 36245–36252.
76. Kumar, M.K.P., Singh, M.P., and Srivastava, C. (2015) *RSC Adv.*, **5**, 25603–25608.
77. Berlia, R., Kumar, M.K.P., and Srivastava, C. (2015) *RSC Adv.*, **5**, 71413–71418.
78. Liu, C., Su, F., and Liang, J. (2015) *Appl. Surf. Sci.*, **351**, 889–896.
79. Yang, Z., Sun, W., Wang, L., Li, S., Zhu, T., and Liu, G. (2015) *Corros. Sci.*, **103**, 312–318
80. Sun, W., Wang, L., Wu, T., Pan, Y., and Liu, G. (2014) *Carbon*, **79**, 605–614.
81. Khalil, M.W., Salah Eldin, T.A., Hassan, H.B., El-Sayed, K., and Abdel Hamid, Z. (2015) *Surf. Coat. Tech.*, **275**, 98–111.

7
Graphene Market Review

Marko Spasenovic

7.1
Introduction

The first production of graphene in the Geim laboratory in 2004 [1], with the now famous scotch-tape method, locked the interest of a large body of scientific researchers. Nearly all of graphene's spectacular predicted properties have since been confirmed experimentally. The scotch-tape, or exfoliation method, yields graphene of a very high quality, but of a small size, with an area generally on the order of a few tens of micrometers squared. One would typically be lucky to find several flakes of that size on a substrate that is 1 cm^2. The extremely low yield of mechanical exfoliation has not been a problem for testing fundamental science; however, it would be too small for any conceivable application.

Already in 2006, researchers succeeded in growing large-area graphene by chemical vapor deposition (CVD) [2], the first method of mass-producing graphene. CVD yields almost 100% coverage of metal films with monolayer graphene. The method of CVD growth on metal and subsequent transfer of graphene to various useful substrates has rapidly overcome many hurdles [3, 4], recently yielding a continuous film of graphene on a 100 m roll [5]. The growing number of graphene researchers coupled with the wide accessibility of CVD growth chambers quickly led to a blossoming of the number of companies marketing graphene sheets. The end user has for the most part been the lab researcher, with almost no excursion to industrial space. The primary reason for the relatively slow inclusion of industry has of course been the price of graphene, which is still higher than that of silicon, and orders of magnitude higher than the price of competing transparent conductor material, indium tin oxide (ITO) [6]. Since graphene-based devices could not beat the price of devices based on more mature materials, the young graphene industry has been looking for those applications where graphene can offer novel benefits or unprecedented functionality. In more recent years, graphene obtained with chemical methods, such as liquid phase exfoliation [7] or the modified Hummers method, has become an instant hit with lab researchers as well as a myriad of companies. Although chemically exfoliated graphene is of lower quality than sheet graphene,

Graphene Technology: From Laboratory to Fabrication,
First Edition. Edited by Soroush Nazarpour and Stephen R. Waite.

it is cheaper to produce and is readily obtained in large volumes [8]. Graphene obtained with these methods is typically in the form of a powder, which means it can be used as an additive in composites, opening a door to a market of a size much larger than the graphene research market [9], spreading across the automotive, sporting, construction, aeronautic, and other traditional large industries, but also affecting emerging technologies such as 3D printing, wearable clothing, and printed electronics. This most recent development pushed graphene closer to real-world applications and ignited a new direction for the market to expand into, enabling a step-up in production volumes and a further reduction of the price of graphene. In addition to few-layer graphene (also dubbed as "graphene nanoplatelets" (GNPs)) some chemical methods yield carbon nanostructures of various shapes. The wide variety of available chemistry results in as wide a variety of graphene-related products, in fact an entire family of "graphenes." There are thus almost as many variants of graphene as there are graphene producers, setting standardization as priority number one for the coming years, but also creating confusion and adding to the difficulty of predicting any future market volumes. One fact is certain: graphene is a disruptive technology that holds enormous potential for commercialization, the realization of which has just begun. It is also an enabling technology, that will infiltrate other key sectors such as 3D printing, healthcare, automotive, aircraft, sports, construction, and others.

This chapter will cover the historical development of the early graphene market (Section 7.2), the current state of the market (Section 7.3), and some future projections (Section 7.4).

7.2 Graphene Market: Past and Present

The early graphene market (2009–2012) revolved around feeding the appetite of the rapidly growing graphene research community. Probably the first company to offer large area graphene in its product catalog was Graphene Supermarket [10], established in 2009 in Calverton, NY. The Supermarket was the sales outpost of Graphene Laboratories, a research start-up run by a few scientists turned entrepreneur. Graphene Supermarket offered CVD-grown graphene on copper substrates and on the most commonly used SiO2/Si substrate, which allows tuning the carrier concentration by back gating. The initial offering was basic, the only other item in the catalog being clean SiO_2/Si wafers for customers to do their own exfoliation; however, due to the exploding demand and the Supermarket's ability to deliver high-quality graphene in a hassle-free manner, the recipe worked like a charm. This simple yet effective recipe soon spread to Europe and the Far East, with the establishment of Graphenea (Spain) in 2010 [11] and Graphene Square (South Korea) in early 2012 [12]. These three managed to quickly master CVD growth and establish themselves as the default graphene suppliers across three continents, setting any latecomers in a difficult position of having to invent advanced applications for an immature material. Still, with the low entry cost of CVD chambers (of the order of USD100k) and quick maturing of the technology of CVD growth, even these three companies had to keep inventing to alleviate the effect of research labs and universities doing their own growth.

In fact, in the short term of the early graphene market, it turned out to be more profitable to provide CVD chambers themselves than the graphene made in those chambers, as exemplified by the success of planarTECH [13]. planarTECH started in 2012 to sell CVD chambers specially tuned for graphene, and already in 2014 surpassed USD1 million in revenue (the entire graphene market turnover in 2013 was 6.6 million EUR [14], according to some, and 20 million EUR, according to others (IDTechEx presentations)). planarTECH had aimed to double their revenue in 2015 (J. Patrick Frantz, private communication). Since similar chambers can be used to grow other monolayer materials such as transition metal dichalcogenides, which are gaining popularity, marketing CVD machines should prove to be solid business in the medium term as well.

In the long term, however, graphene holds enormous potential (see Section 10.3), but the trick for success lies in recognizing the correct applications for each variant of graphene and aligning to the needs of the target customer accordingly. The two oldest graphene producers, Graphene Laboratories and Graphenea are proving to be extremely versatile and inventive in that regard, having significantly expanded their product catalogs and engaged in various collaborations with other industries. Since these two are excellent examples of successful companies in the early graphene market, we will focus more on their business strategies before moving to examples of smaller companies. As of date, Graphenea's main focus is still on the research community, although an increasing percentage of business is coming from sales to companies (Jesus de la Fuente, private communication). With the number of graphene-related scientific publications per year soaring in the thousands, this might be a good strategy; however, Graphenea will have to keep following the brutally fast research developments to stay on top. What the company has done so far is to perfect the process of transfer of graphene from copper to custom substrates, offering graphene on PET, glass, SiO_2/Si, TEM grids, and microcavities as standard in the catalog, with a promise to be able to transfer on almost any substrate that a customer provides.

Following the developments related to powdered forms of graphene, Graphenea has introduced various products featuring graphene oxide, either in a water suspension or as dry powder. These products are in demand, as powdered graphene is easily integrated into polymer matrices for enhancement of thermal and mechanical properties. It will most likely be powdered graphene (including GO) that keeps Graphenea sailing in the medium term, as the construction, automotive, aerospace, and sporting industries pick up on the hi-tech functionality that graphene can offer. At the end of 2013, Graphenea announced a million-euro investment in the company by Repsol [14], a leading oil and gas business. Graphenea has managed to increase its turnover by 50% each year since establishment. To keep on top of cutting edge research, Graphenea collaborates with leading graphene scientists, which gives the company an excellent knowledge base as well as a good reputation. Graphenea is also the leading graphene supplier in the billion-euro Graphene Flagship consortium, a 10-year program of the European Commission designed to help bring graphene to commercial products. Becoming part of the Flagship provided access to hundreds of European research groups for Graphenea, further strengthening the company's position on the European market. The business strategy for Graphenea is to follow the evolution of graphene as it moves from basic research to applications, and in that regard it is natural that the user base has been gradually shifting from research labs to other businesses in end-user industries. In 2014, the company had more than USD1 million in revenue, and a positive cash flow. The premise is that the price of high quality sheet graphene will soon drop below the price of main competing materials (see Figure 7.1), which will give a boost to sales of pure graphene.

Graphene Laboratories has taken a different approach by expanding their product catalog to include other ultrathin materials such as boron nitride, MoS2, WS2, as well as materials and devices supplementary to graphene production, like coronene for better CVD growth or the SpeedMixer for mixing nanomaterials and creating paste.

In February 2012, Graphene Labs signed an agreement with Lomiko Metals Inc, a Canada-based mining company, to build entire graphene application chains. The agreement obliged Lomiko to provide high-quality natural graphite and to raise USD2 million over a period of 18 months. In effect, the companies worked together to develop graphene technology and to perfect the use of natural graphite in graphene production.

This type of joint venture between graphene specialists and graphite mines was also seen elsewhere, for example in the partnerships of Grafoid and Focus

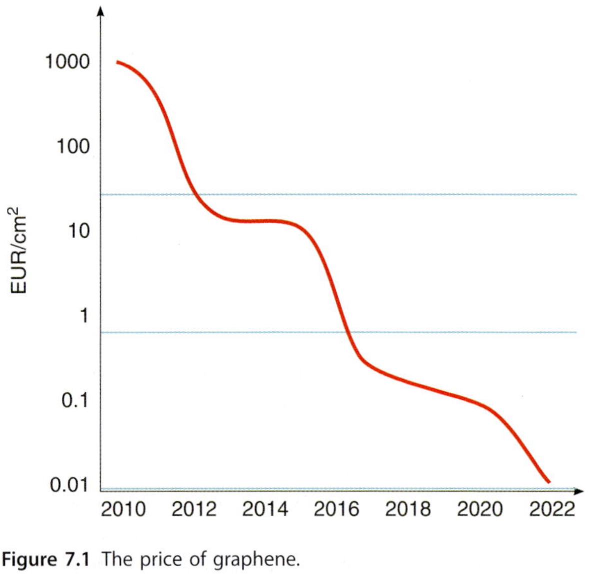

Figure 7.1 The price of graphene.

Graphite, Mason Graphite and Group Nanoxplore, and American Graphite and Cheaptubes. Although it seemed at the time that many more graphite miners would go this way, others seem to have shied away from major graphene ventures. For example, Northern Graphite, another large mining company based in Canada, was happy to sign a research agreement with Turkish Grafen Chemical Industries, with the only stake being patent rights. Overall, the graphite mining industry has been an important investor in the graphene market for the past 5 years, mainly through joint ventures with and takeovers of graphene specialist firms. In this period, graphite mines were one of the very few options for investing in "graphene stock," albeit rather indirectly. By now several pure graphene companies have gone public, making the graphene market more interesting to smalltime investors.

Graphene Labs have used their presence on the ultrathin materials market as a springboard to hop into 3D printing, a fast-emerging tech that relies on novel materials where graphene would feature as an additive to the printing mass, either as conductive material or as a mechanical strengthener. The founders of Graphene Labs spun out Graphene 3D Lab Inc., and in 2014 demonstrated a 3D printed graphene-based battery, that can be printed in any shape and size and can therefore be incorporated into custom designs. At the end of that year the company acquired Boots Industries, a Canadian 3D printer manufacturer. Following up, the company has launched a conductive graphene-enhanced filament, fully compatible with commercial 3D printers. A conductive 3D printing filament can be used, for example, to print EMI and RF shielding, wearable electronics, keyboards, MIDI controllers, and so on.

In June 2015, Graphene 3D Labs announced the doubling of their production capacity for functional filaments, triggered by growing demand on their graphene filament. The company also announced the introduction of at least three new materials in 2015. Clearly, this is a successful line of business with huge growth potential, and Graphene Labs has been playing it right from the onset. Like Graphenea, Graphene Laboratories announced revenue in excess of USD1 million in 2014, and a positive cash flow coupled with a lack of debt.

Perhaps the biggest producer (west of China) of GNPs, tiny flakes of multilayer graphene, is Dayton based Angstron Materials. Claiming approximately 300 tpa production capacity and several dozen patents on GNPs, this is also the company that holds the first patent on GNPs. "GNPs" is a generic name given to a multitude of graphene nanomaterials of arbitrary shape and lateral dimensions on the order of 100 nm, with a thickness of several nanometers. This kind of graphene has been gaining popularity in recent years, due to the ease of production and the chemical reactiveness at the edges, which makes it suited for integration into other material matrices. GNPs are heavily being used in thermal, mechanical, barrier, and energy applications.

Most producers have been moving up the value chain, offering additive masterbatches, inks and coatings, and partnering with other industries to push graphene into products. For example, XG Sciences have delved deep into energy storage applications, partnering with Samsung Ventures Investment Corporation and having already tested graphene-based batteries and electrodes. Vorbeck has gone even further, offering ready-made RFID tags, enhanced rubber, and flexible battery straps made of graphene. Similarly, in parallel to sales of various forms of GNPs, Italian Directa Plus partnered with Vittoria SpA to bring to the market the world's first graphene-enhanced bicycle tires, with the next step going in the direction of race car pneumatics. The sports requisite industry in general provides a natural habitat for graphene products, because sports equipment needs to be lightweight, durable, and often flexible, all of which can be improved with the addition of a little bit of graphene to the original polymer matrix. In that respect, the first commercial product containing graphene was a tennis racquet from HEAD, sported already in 2013 by Novak Djoković and Maria Sharapova.

Although several companies claim hundreds of tpa capacity and a few claim kilotonnes, it is estimated that no more than a kilotonne per year of combined production of GNPs and graphene oxide will be required in the next few years [15], indicating current production overcapacity. Graphene manufacturers are hence now focusing on batch-to-batch consistency and cutting production costs [15]. Overproduction is best illustrated by the resultant effect on stock of some publicly traded graphene businesses, shown in Figure 7.2.

It is likely that the graphene market has just passed the peak of hype and is entering the trough of disillusionment of the Gartner hype cycle (Dr Khasha Ghafferzadeh, IDTechEx presentations). However, some have managed to have short-term stock market success, as exemplified by the UK-based Haydale.

The market capitalization for all pure-play graphene companies is well below USD50 million, setting them in the "nano cap" category. It seems that as of the end of 2015, the graphene market is at a tipping point between fundamental research-driven and application-driven (Figure 7.3).

The graphene market and its key players are best summarized in Figure 7.4 where development of graphene is tracked from source to end product through the value chain. Although only key businesses are shown on the chart, the Graphene Tracker business directory contains nearly 90 entries, and there are an estimated 44 graphene producers on the market [15]. Most fit in the "Graphene Based Compounds" category, and at least half of those are moving up the value chain, with their own end products in the pipeline. Due to the oversupply and rapid growth of the number of companies in this sector, moving up the value chain seems like the only good strategy for long-term survival. Also, a common strategy for graphene businesses has been to find partners further up the value chain, in a joint effort to propel graphene materials to higher value. Examples of this type of partnership abound, for instance in Angstron's ventures with Stryke Industries (military), Haydale's with planarTECH (graphene manufacturing equipment), Graphene 3D Lab's with Polymaker (3D printing materials), XG Sciences' with Samsung (electronics), Perpetuus' with OXIS (battery electrodes), and Graphene NanoChem's with Emery (oleochemicals).

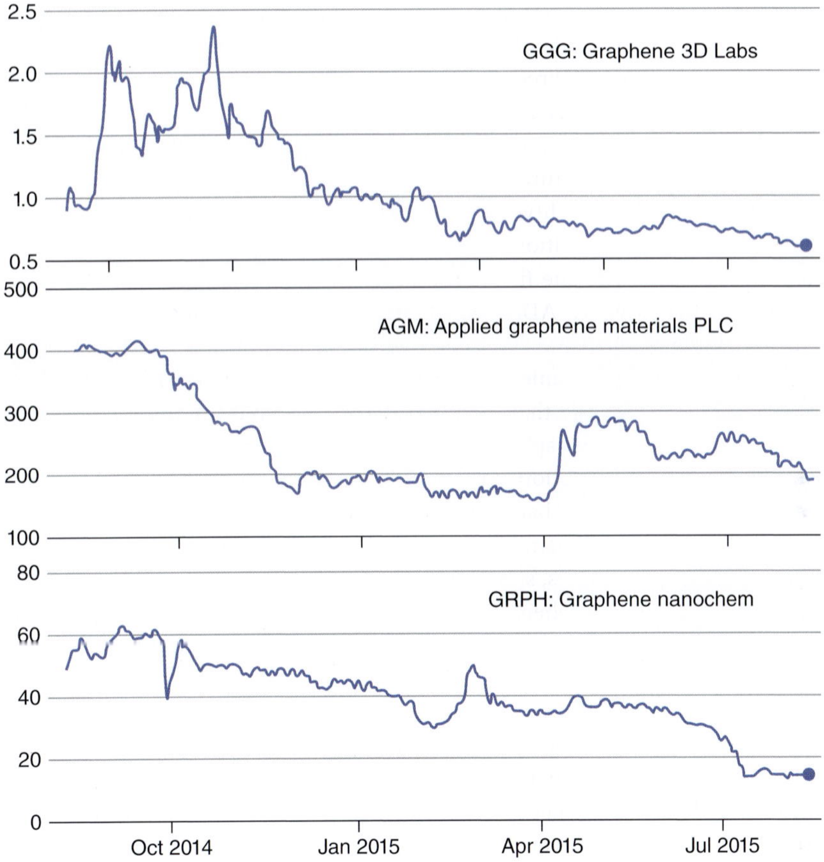

Figure 7.2 Stock value of some publicly traded graphene companies (Google Finance.)

Figure 7.3 Haydale stock price.

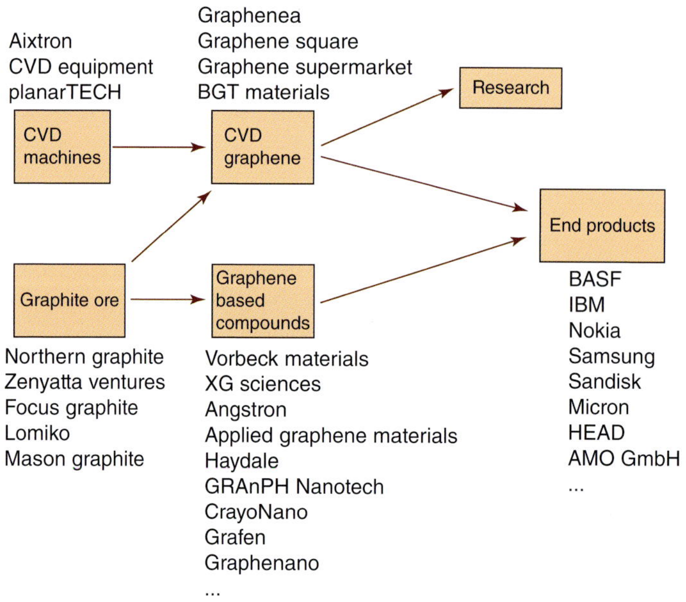

Figure 7.4 Graphene industry diagram, highlighting key businesses. (www.graphenetracker.com.)

7.3 Co-ordinated Market Initiatives

In an unprecedented effort to turn graphene expertise into market value, large funding and coordination initiatives are being assembled worldwide. By far, the largest of these is the pan-European Graphene Flagship, which aims to coordinate research and commercialization efforts of 142 entities in 23 countries, as of March 2015. The Flagship is both a funding and coordination action, with the EU contribution amounting to 1 billion EUR over 10 years (2012–2022) and as much expected to be contributed by national governments and businesses. The Graphene Flagship is run from Chalmers University of Technology in Sweden, directed by Prof. Jari Kinaret. The director is supported by a management team, and the entire undertaking is divided into 16 work packages, each coordinated by its own leader. The Flagship is in close contact with the European Commission, the executive body of the European Union, and the two have signed a partnership agreement that obliges them to support each other. Regional, national, and transnational graphene-related projects can apply to become part of the Graphene Flagship, which allows for absorbing the entire graphene community of Europe into one body.

The Flagship is one of two Future and Emerging Technology (FET) initiatives (the other being the Human Brain Project), the success of which will decide whether the EU will embark on other such long-term projects in the future.

On the North-American side, the effort is not as co-ordinated as the Flagship consortium, but it rather reflects a traditional American approach. The commercialization effort is being driven by the Graphene Stakeholders Association (GSA), a non-profit organization entailing more than 40 members. The organization pushes forward standardization, characterization analysis, and environmental health and safety issues, in an effort to legislate graphene. Through on-going collaboration with key stakeholders, the GSA's "Center of Excellence" initiative seeks to accelerate the development of standards and facilitate the commercialization process of graphene and other 2D materials. In the far East, the China Innovation Alliance of the Graphene Industry (CGIA) is the Chinese counterpart of the GSA. Also a non-profit, this alliance is a national organization supported by the government of China. The CGIA is a consortium of several dozen members, including industrial enterprises, academic institutions, and research organizations. The South Korean government announced a USD40 million 6-year investment plan for graphene companies in 2013. Finally, Malaysia announced the Malaysian "National Graphene Action Plan" in July 2014. A "white paper" was published at the time, but little news followed since.

More local intensified efforts to commercialize graphene are evident in the United Kingdom, where government support to bridging research and industry has materialized through the National Graphene Institute (NGI) and the Cambridge Graphene Centre. The Cambridge Centre was initially funded by a £12 million government grant, coupled with an additional £13 million from industrial partners, including Nokia, Plastic Logic, Dyson, Philips, and BAE Systems. The NGI in Manchester is a research and incubator center housed in a £61 million building of over 7600 m^2. The first company to move into the NGI was BGT Materials, formerly known as Bluestone Global Tech.

7.4 Market and Application Projections

Projections of the value of the graphene market vary almost as much as the forms of graphene. Most recent reports from IDTechEx predict a market value of about USD200 million in 2026, and include 17 specific application sectors. The largest application sectors, according to this report, will be energy storage and supercapacitors, composites, and research, with inks and coatings following closely.

In contrast, a report from Chisult Insight [16] predicts a USD300 million market size for 2020 in China only. Allied Market Research predicts USD150 million globally for the same year; Markets and Markets predicts USD1.5 billion in 2020 from electronics only [17], while BCC Research predicts about USD1 billion in total [18]. Experienced market analysts, including IDTechEx and Lux Research, are increasingly cautious about their predictions, with Lux Research's publishing an extremely negative outlook in June 2015 [19].

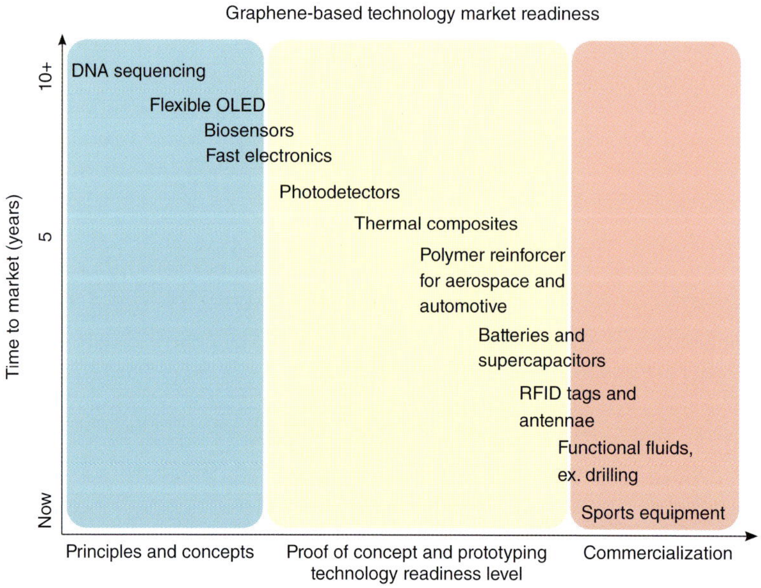

Figure 7.5 Time to market and technology readiness level of some applications (author's own predictions).

Applications that demand fast electronic performance, which were first thought to be the most promising ones for graphene, will surprisingly grow very little in the short term, due to the fierce competition that graphene faces there from established materials such as silicon and ITO. In general, applications that require high-quality graphene are expected to take longer to reach the market, as evidenced by recent application predictions (see Figure 7.5 and Refs [20] and [21]). IDTechEx predicts that even in the year 2026, more than 75% of the market share will belong to composites, energy storage (supercapacitors), and inks and coatings, whereas sensors, logic, transparent conductive films and research will altogether take up less than 15% of the market share [21].).

7.5 Conclusion

Although still primarily research-driven, the graphene market is naturally being populated by graphene for industrial use. The effort to commercialization is especially being helped by large co-ordinated initiatives on all sides of the globe. Drawing from the experience with carbon nanotubes and other nanotechnologies, the community is pursuing material substitution and additives in parallel to discovering disruptive, groundbreaking applications. The graphene production market is following controlled expansion and is now worth several million USD, with projected rapid expansion in the coming years due to the material's diverse range of applications.

References

1. Novoselov, K.S., Geim, A.K., Morozov, S.V., Jiang, D., Zhang, Y., Dubonos, S.V., Grigorieva, I.V., and Firsov, A.A. (2004) Electric field effect in atomically thin carbon films. *Science*, **306** (5696), 666–669.
2. Somani, P.R., Somani, S.P., and Umeno, M. (2006) Planer nano-graphenes from camphor by CVD. *Chem. Phys. Lett.*, **430** (1), 56–59.
3. Li, X., Cai, W., An, J., Kim, S., Nah, J., Yang, D., Piner, R., Velamakanni, A., Jung, I., Tutuc, E., Banerjee, S.K., Colombo, L., and Ruoff, R.S. (2009) Large-area synthesis of high-quality and uniform graphene films on copper foils. *Science*, **324** (5932), 1312–1314.
4. Bae, S., Kim, H., Lee, Y., Xu, X., Park, J.-S., Zheng, Y., Balakrishnan, J., Lei, T., Kim, H.R., Song, Y.I., Kim, Y., Kim, K.S., Özyilmaz, B., Ahn, J.H., Hong, B.H., and Iijima, S. (2010) Roll-to-roll production of 30-inch graphene films for transparent electrodes. *Nat. Nanotechnol.*, **5** (8), 574–578.
5. Kobayashi, T., Bando, M., Kimura, N., Shimizu, K., Kadono, K., Umezu, N., Miyahara, K., Hayazaki, S., Nagai, S., Mizuguchi, Y., Murakami, Y., and Hobara, D. (2013) Production of a 100-m-long high-quality graphene transparent conductive film by roll-to-roll chemical vapor deposition and transfer process. *Appl. Phys. Lett.*, **102** (2), 023112.
6. www.graphenea.com/pages/graphene-price (accessed 13 January 2015).
7. Du, W., Jiang, X., and Zhu, L. (2013) From graphite to graphene: direct liquid-phase exfoliation of graphite to produce single- and few-layered pristine graphene. *J. Mater. Chem. A*, **1**, 10592.
8. Wolf, E.L. (2014) Chapter 2: Practical productions of graphene, supply and cost. Applications of Graphene, SpringerBriefs in Materials.
9. www.lucintel.com (accessed 13 January 2015).
10. graphene-supermarket.com (accessed 13 January 2015).
11. www.graphenea.com (accessed 13 January 2015).
12. www.graphenesq.com (accessed 13 January 2015).
13. www.planartech.com (accessed 13 January 2015).
14. *Repsol Press Release*, http://www.repsol.com/es_en/corporacion/prensa/notas-de-prensa/ultimasnotas/05122013-repsol-cdti-invierten-un-millon-euros-empresa-graphenea.aspx (accessed 13 January 2015).
15. Zurutuza, A. and Marinelli, C. (2014) *Nat. Nanotechnol.*, **9**, 730–734.
16. http://www.reportsnreports.com/reports/402474-research-on-global-china-graphene-industry-2015.html (accessed 29 February 2016).
17. http://www.researchandmarkets.com/publication/mystl29/graphene electronics_market_by_materials_dev (accessed 29 February 2016).
18. http://www.bccresearch.com/market-research/advanced-materials/graphene-technologiesapplications-markets-avm075c.html (accessed 29 February 2016).
19. http://blog.luxresearchinc.com/blog/2015/07/case-closed-graphene-is-the-next-carbon-nanotube/ (accessed 29 February 2016).
20. Ferrari, A.C. *et al.* (2015) Science and technology roadmap for graphene, related two-dimensional crystals, and hybrid systems. *Nanoscale*, **7**, 4598–4810.
21. http://www.idtechex.com/research/reports/graphene-2d-materials-and-carbon-nanotubes-markets-technologies-and-opportunities-2015-2025-000440.asp (accessed 02 May 2016).

8
Financing Graphene Ventures

*Stephen R. Waite**

> Graphene is dead. Long live graphene.
>
> *-Andre Geim*

The discovery of graphene in 2004 has generated a good deal of interest among investors, entrepreneurs, and established companies alike. Graphene is viewed today as one of the world's most remarkable materials and for some very good reasons. Graphene possesses unique properties (as discussed elsewhere in this book) that make it a promising candidate for use in a range of applications in key economic segments, including aerospace, consumer products, energy, electronics, communications, health care, and transportation. The astonishing properties of graphene have captivated the interest of many investors and entrepreneurs who see the potential to add significant value in commercial applications using the material.

Over the past decade, numerous companies have been started to capitalize on the opportunities associated with graphene. While these companies vary in their purpose and objective, they all share a common trait in needing to raise capital to fund operations, research, and development. In this chapter, we explore the issues related to financing companies innovating with graphene.[1)]

This chapter comprises five major sections plus an Appendix. The first section provides a brief overview of the graphene company landscape. The second section explores what we call "The Art of Raising Capital," in both theory and practice. Raising capital is paramount for all graphene ventures and this section provides a primer on funding channels for graphene ventures. The third section examines the shifting financial landscape and discusses the implications for seed and early stage graphene ventures. This section includes observations gleaned from interviews with various executives in regard to raising capital for their graphene ventures in the current financing environment. The fourth section discusses the financing

* Co-founder and Co-Executive Director of the Graphene Stakeholders Association, author of *Quantum Investing* (Thomson/Texere).

1) It should be noted and kept in mind that while the focus of this chapter is on financing graphene ventures, many of the observations apply to seed and early stage companies innovating with other 2D and 3D advanced nanomaterials.

Graphene Technology: From Laboratory to Fabrication,
First Edition. Edited by Soroush Nazarpour and Stephen R. Waite.

road ahead for graphene ventures in light of technological and financial market trends. The fifth section provides a summary of key points. The Appendix includes a case study of Nantero, Inc., a company pioneering the development of innovative memory devices using nanomaterials. The case study provides a perspective on the opportunities and challenges of funding start-ups engaged with commercializing innovative products with advanced nanomaterials in the twenty-first century.

8.1 Graphene Start-ups

The number of start-up companies in the graphene space mushroomed after 2010 when the discoverers of the material, Geim and Novoselov at the University of Manchester in the United Kingdom, were awarded the Nobel Prize. One can classify graphene start-ups into three broad categories:

1) Producers and suppliers of graphene and graphene derivatives such as graphene nanoplatelets (GNPs) and graphene oxide (GO); this group also includes companies focused on functionalizing graphene for various commercial applications;
2) Companies developing graphene-based applications. In this chapter, we will use the term graphene generally to encompass derivatives such as GNPs and GOs. Moreover, it should be noted that the following discussion pertains to companies innovating with other 2D and 3D advanced nanomaterials – those materials that have been discovered and those to be discovered in the future.
3) Metrology and tools manufacturers.

With respect to the first group, there have been numerous start-ups engaged in the production and supply of graphene. There are several methods to produce graphene, including chemical vapor deposition (CVD) and exfoliation. There is great interest in developing scalable production methods that can significantly boost the quantity of graphene in the market while simultaneously driving down the cost. Emerging companies in this segment include Graphenea in Spain, Applied Graphene Materials in the United Kingdom, XG Sciences in America, Graphene Platform Corp. in Japan, 2D Carbon Tech in China, and Poland's Advanced Graphene Products. Graphenea and XG Sciences are also engaged in developing various graphene-based applications so there is an overlap with the second classification of companies noted above.

Also included in the first group of graphene ventures are companies focused on functionalizing the material for use with other materials in various applications. Functionalization of graphene provides potential users with a way to tailor the materials for use in composites. Companies pioneering the functionalization of graphene include Haydale PLC in the United Kingdom and Group NanoXplore in Canada. Group NanoXplore is also a supplier of graphene so there is some overlap with the producers and suppliers.

The second segment of start-ups is focused on using graphene in next-generation applications in a range of industry segments. Graphene Frontiers in the United States, for example, is focused on graphene-enabled biosensors for use in the health care sector. US-based Bess-Tech is focused on next-generation graphene-enabled battery technology; Graphene Batteries, based in Norway, is developing safe and durable graphene- based high-energy battery materials. San Francisco-based Graphene Technologies has developed a process to turn CO_2 into graphene for a range of applications. Graphene Nanochem is a UK-based company that manufactures performance specialty chemicals and advanced nanomaterials from renewable sources including waste materials. Graphene Lighting PLC is a spin-out company based on a strategic partnership with the National Graphene Institute (NGI) at The University of Manchester to create graphene applications. NanoPhyll is a Canada-based private company dedicated to improving efficiencies in the photocatalytic and photovoltaic markets through the application of customized graphene. Calevia is a Canadian start-up company co-developing a graphene-based cancer thermal treatment platform. Two Carbon is an advanced materials technology company building high-performance product solutions for the outdoor consumer.

Additionally, a growing number of fledging graphene companies are engaged in using graphene for use in 3D printing and additive manufacturing, an area that is pregnant with innovative possibilities and opportunities for value creations. Companies focused in this area include Graphene 3D Labs and 3D Graphtech Industries.

The third group involves manufacturers and suppliers of metrology and tools essential for the production of graphene and graphene-enabled applications. Companies in this segment include, Veeco Instruments, PlanarTech, Aixtron, Annealsys, CVD Equipment, Moorfield, and Picosun. As a group, these companies are, generally speaking, more established and have been serving customers for many years. Graphene and other advanced 2D nanomaterials represent a new and potentially lucrative market segment for them.

As can be seen, there are numerous early stage companies working to commercialize many different types of graphene-enabled products in many countries and regions around the world. The varied activities associated with graphene are global in nature, but graphene is not an industry. Rather, graphene is an advanced nanomaterial with remarkable properties that can be harnessed and used to innovate in a wide variety of industries for numerous applications – some of which are potentially game changing. The nature of graphene makes it an attractive investment opportunity. We will continue with the second section of this chapter and a discussion of the art of raising capital for graphene ventures.

8.2 The Art of Raising Capital

Deep science lies behind the commercialization of graphene, but there is no science behind the financing of companies innovating with advanced nanomaterials. Financing start-ups is largely an art. There is a great deal of trial and error involved in funding early stage companies and failure to complete a funding round is an all-too-often common experience. When it comes to financing a start-up, it is through failure at raising funds that one eventually succeeds. Nothing succeeds quite like failure when it comes to raising capital for a seed or early stage technology firm. The amount of "no thanks" or "nope, not interested" a start-up founder and his/her team encounters during capital raising activities far outweighs the "yeses" one gets. This is important to keep in mind during the seed stage of a start-up. Stay positive at every juncture, but realize that the process of raising capital can, and often is, a frustrating activity that can tie up a great deal of time.

Raising capital is an art. One would undoubtedly prefer to have fund raising for an early stage venture to be more like a sprint, but the experience of capital raising shows that it is more akin to a marathon. Many founders and CEOs spend a great deal of time raising capital – time that can and often does distract from running the day-to-day affairs of the business. Financial market conditions are constantly shifting – often in unpredictable ways. Changing conditions in financial markets often produce shifts in investor appetite and valuations – sometimes in a dramatic fashion that are beyond the control of the company and the executive(s) tasked with raising capital.

Stock markets and interest rates can bounce around like yo-yos complicating the closing of an early or even late stage round of funding. There are times when the equity markets are trending up and investors are bullish gobbling up initial public offerings (IPOs) and pouring billions of dollars into private equity (PE). Then comes along some event or series of events that results in a dramatic shift in the financial environment where the IPO window closes and funding dries up. We have experienced times when emerging technology companies are ready to close the books on an IPO only to see the IPO window close and not open back up for many months. The entrepreneur raising capital has no control over the financial markets and IPO window. He or she is at the whim of Mr. Market.

The harsh truth is that founders and executives of start-up companies do not have a crystal ball and cannot accurately foresee the coming swings in financial markets – swings that directly impact the ease or difficulty of raising capital. It is the unpredictable nature of Mr. Market that makes raising capital for start-ups an art as opposed to a science. To be sure, the art of raising capital for start-ups is augmented by experience and success with previous start-ups. This helps explain why successful entrepreneurs with proven start-up company success are prone to have a relatively easier time raising capital than other less proven entrepreneurs.

Entrepreneurs starting up new ventures typically seek out multiple funding sources, and for good reason. While there is often a wish list of what type and size of funding one would like to have, it is often the case that entrepreneurs have to take what the market gives them. When it comes to raising capital for a start-up, there is no substitute for developing and having a diverse network of potential investors from which to source capital. A diverse network of potential investors gives the founder or CEO of a seed or early stage company greater access to capital. As a general rule of thumb, the more sources of capital, the easier and faster it is to raise capital. There is no one single source of capital that a start-up can rely on to adequately fund its operations. Potential sources of capital for start-ups focused on graphene ventures include the following:

1) Founder, Friends, and Family (F&F)
2) High Net Worth private investors (HNW)
3) Angel investors (Angels)
4) Venture Capitalists (VCs)
5) Private equity investors (PE)
6) Government funding – Federal, State, and Local (G)
7) Strategic investors – corporations (SI)
8) Reverse Merger
9) IPO

As can be seen from this list, there is a variety of sources of funding for graphene start-ups and early stage companies developing innovative products with graphene. We will discuss each of these segments with a focus on the role each play. These nine segments of funding sources can be grouped into two broad categories: Retail and Institutional. There is some blurring among these categories in this day and age of Angel networks and crowdfunding, which are discussed further. F&F money falls clearly into the retail group of investors, while VCs, PE and SIs are classified as institutional investors. HNW and Angel investors are a mixed breed of retail and institutional. Additionally, there are institutional sources of capital that rely on retail money. IPOs fall into this category. Institutional investors such as investment banks lead manage IPOs but tap into pools of retail and institutional money to complete the offerings. We will come back to IPOs and discuss this potential source of growth capital for graphene ventures.

The classification of investor groups can be further broken down into passive and active. Passive investors provide funding but have no role in developing the company, while active investors are engaged in various roles to help foster growth and value creation. Generally speaking, retail investors are passive while some institutional investors are active in supporting the companies in which they invest. A percentage of institutional venture capital firms actively invest in early stage companies. Start-ups generally rely on a mix of retail and institutional funding, and the graphene (and advanced nanomaterials) segment is no exception. The lines between these two categories are getting blurred with the development of Angel networks, which have emerged as a viable funding source for graphene ventures in the past several years.

In the start-up and early phases of operation, the lion's share of funding typically comes from founders and F&F. The amount of money raised through this source varies depending on one's personal net worth and the net worth of friends, family, and community. Typically, the founder will invest a certain amount of money to fund the legal entity and begin operations. F&F contributions augment the founder's investment and allow for further growth in intellectual property development and protection and the hiring of staff, advisors, and various business support services.

It is noteworthy that the advent of social media has found an application in the realm of fund raising in recent years with crowdfunding (or crowdsourcing) platforms such as Kickstarter and others. These social media–based platforms are altering the way start-up capital can be raised. Crowdfunding has gained traction with entrepreneurs, as it can be a quick and efficient way to raise seed capital from a diverse group of retail investors that extends beyond the traditional F&F channel. While crowdfunding rules are different across countries, they all employ social media to raise capital. What's more, while crowdfunding has been more retail in nature in its early development, more institutional-type crowdfunding platforms have been emerging in the market (e.g., SeedInvest, CircleUp). The emergence of institutional crowdsourcing platforms could become an important future source of funding for start-ups engaged in developing graphene-based innovations as time goes on.

Company Financings At a Glance

Seed Financing: provided to research, assess, and develop an initial concept before a business has reached start-up phase; customer development is key during this phase.

Early Stage Financing: for product development and initial sales and marketing.

Expansion/Growth Financing: for growth and expansion of the company that is selling products and seeking to penetrate more deeply into the marketplace. Capital may be used to finance increased production capacity, market or product development, and/or to provide additional working capital.

Replacement Financing: Purchase of shares from another investor or to reduce leverage via the refinancing of debt.

Buy Out Financing: The acquisition of a significant portion, majority control or 100% of businesses that normally entail a change of ownership. Funds are often used for expansion, consolidations, turnarounds, and spinouts of divisions or subsidiaries.

Another prominent and promising source of capital for seed and early stage graphene ventures is HNW investors. Such investors can be a valuable source of early stage capital for graphene ventures. HNW investors encompass a wide variety of people, many of whom have enjoyed success in business and are eager and willing to put their hard earned money to work in a risky venture with promising investment returns. Many HNW investors are local, but have extensive networks that extend nationally and globally and include a variety of experienced people in services such as legal, financial, and operational – services that are essential for early stage companies. A simple phone call by an HNW may result in finding a valuable resource for the company in a matter of a day or two that otherwise may take many weeks or months to find. Some HNW investors have Board of Director and/or a high level of operational experience that can be extremely helpful to an early stage company. Finding an HNW investor or several HNW investors with access to millions of dollars of capital can make a difference in the early life of an emerging technology venture working with graphene and other advanced nanomaterials. Some HNW investors also act as angel investors and participate in venture capital funds, as discussed further later.

Angel investors have emerged in recent years to become another viable and important source of capital for seed and early stage technology companies. Angels typically play the role of a bridge between F&F and Institutional sources of capital such as venture capital. Angel investors can be similar in nature to HNW investors in being local and experienced in the ways of evaluating businesses and having access to funding that extends well beyond the capabilities of F&F. Some Angels are also HNW investors, thus blurring the line between the two groups. There are also Angel investor networks that are regional and, as such, have access to larger amounts of capital. Such Angel networks, which can be found in various large metropolitan areas in the United States and elsewhere, operate more like an institutional investor akin to a VC. Angel funding can be, and sometimes is, a stepping-stone to venture capital funding, should a company seek to pursue such financing if it is available.

Venture capital is the most organized type of PE financing available and a source of capital that has a distinguished history of funding emerging technology companies involved with commercializing advanced materials. The venture capital business that has evolved over the past half century has its origins in funding risky emerging technology ventures. VCs have evolved into a large and thriving business geared at financing entrepreneurs and risky seed and early stage technology companies.

Venture capital provides a channel whereby HNW investors can participate in higher risk ventures without having to personally undertake the burden of venture evaluation and management. The VC fund itself provides the expertise in sourcing, evaluating, investing, managing, and harvesting the venture investments. Most Angels invest in their own right, whereas VC investment is through a fund. The common structure of a venture capital fund is the limited partnership (LP). This structure is commonplace in both the United States and United Kingdom as well as in Australia.

Venture capital funds are typically established for a 10-year life, but may be extended in some circumstances. The investor in a VC fund is referred to as a limited partner. The LP commits to make available funds as needed for the underlying venture investments. The investment manager (otherwise called a general partner or GP) organizes and conducts the financings. As investments are harvested, proceeds are returned to the LPs and not reinvested into new opportunities. Since the timing of exits cannot be known in advance, limited partners must be prepared to wait for some time before they start to see any return on their fund's companies.

The structure of a typical venture capital fund, while conducive to funding risky emerging technology companies commercializing advanced nanomaterials such as graphene, also creates incentives for GPs to seek near term liquidity events through IPOs and corporate acquisitions. Many venture capital funds in Silicon

Valley have been migrating away from funding advanced nanomaterials companies due to a variety of factors. The proto-typical seed or early stage social media or software company is seen to offer a more compelling expected risk/return profile to VCs relative to graphene and other advanced nanomaterials ventures. This trend will be further discussed as it has major implications on the funding of graphene ventures today.

Another source of capital for graphene ventures are PE investors. PE is akin to venture capital in that both are traditionally based on the LP structure and invest directly in companies. That said, PE investors are more geared toward later stage financings and many are focused on buyouts – sometimes employing a large amount of leverage through the issuance of debt. PE funds are typically focused on the long-term potential of the portfolio of companies they hold an interest in or acquire. Once they acquire or control interest in a company, PE funds look to improve the company through management changes, streamlining operations or expansion, with the eventual goal of selling the company for a profit, either privately or through an IPO. To achieve their aims, PE funds usually have, in addition to the fund manager, a group of corporate experts who can be assigned to manage the acquired companies. The very nature of their investments requires their more long-term focus, thus distinguishing PE funds from other types of investors that have shorter investment time horizons – horizons that conflict with developing innovative products with graphene and other advanced nanomaterials. That said, PE investors typically do not participate in funding of early stage emerging technology ventures and are more of an option for founders and executives of later stage graphene ventures seeking liquidity or an exit.

The four remaining channels of capital for graphene ventures to be discussed are: government funding, SIs, reverse mergers, and IPOs. We will take each of these in turn.

Government funding has long been an important source of capital for emerging technology companies. There are several reasons for this, chief among them being the demand for innovative technologies related to national defense. There are a host of innovative technologies in a variety of segments that have benefited from government funding, the Internet being a recent prime example. Government funding can be attractive to seed and early stage graphene ventures given that many sources of government funding are aligned with the time horizon of developing innovative technologies with advanced nanomaterials. Government funding tends to be more patient relative to other sources of funding. As such, access to government funding through grants and other initiatives can be extremely beneficial in the early evolution of a graphene venture. Within the government segment, there are a diverse number of channels graphene ventures can source for capital depending on locality, region, and nation. Many governments around the world – particularly in Europe and Asia – are actively and enthusiastically supporting research and development of products and applications using graphene. Government funding is an important source of early stage funding for graphene ventures in these regions.

Another promising channel of funding for graphene ventures is strategic funding from larger, more established corporations. Strategic partnerships with established corporations can provide not only financial capital, but also access to customers as well as helping build an ecosystem to support future development and innovation. Customer development is paramount for success in the marketplace. A well thought out and properly executed strategic partnership with an established corporation can yield great benefits to graphene ventures and set the stage for a liquidity event down the road, through either an acquisition or IPO. Early stage graphene ventures are encouraged to seek out win–win strategic partnerships with established corporations globally as a means of funding operations and developing customers.

The other prominent sources of funding for graphene ventures are reverse mergers and IPOs. Reverse mergers are financing vehicles that supply capital and liquidity for existing and prospective investors and are executed in lieu of an IPO. With a reverse merger, a company backs into a shell that is already publicly traded on a stock exchange case at the OTC market in the United States Some shells have cash on the balance sheet. The cash can be used to help execute a reverse merger. Some investors have a dim view of companies that engage in reverse mergers. There has been quite a bit of financial shenanigans in the form of smoke and mirrors that have gone on with reverse mergers over the years that has tarnished their reputation with investors. That said, a well-executed and managed reverse merger can be a boon to an early stage company developing innovative products with graphene and other advanced nanomaterials.

IPOs are another source of capital for graphene ventures. IPOs are viewed as attractive by many emerging technology companies, primarily because such offerings can and often do provide access to a large amount of capital while creating liquidity for existing and prospective investors. IPOs have a long and illustrious history with emerging technology companies in the United States and elsewhere. Some of the biggest and most successful technology companies, including Apple, Intel, Microsoft, ARM Holdings, Google, Amgen, and many others completed successful IPOs and continue trading on public securities exchanges today.

IPOs has traditionally been viewed as an ideal long-term source of capital given that successful IPOs can take care of present capital requirements and provide access to future capital through secondary and follow on offerings. It is frequently the case that an emerging technology will tap the public markets for additional funding several times early on in its evolution. Financial markets around the world have established stock exchanges where public securities can trade following an IPO. Major banks have extensive resources to help companies execute an IPO and offer research and other banking support following the completion of a deal. A diverse group of investors – retail and institutional – can and do frequently participate in an IPO. There have been a handful of graphene companies that have completed successful IPOs/reverse mergers in recent years, as discussed further.

As can be gleaned from the foregoing discussion, there are many different sources of funding available for graphene and other advanced nanomaterials ventures. The art of financing such ventures involves being opportunistic and resourceful with respect to all the various types of funding sources highlighted earlier. Funding opportunities vary by region and by country. Entrepreneurs seeking to seed or fund early stage graphene ventures are encouraged to explore all channels and sources of available capital in their respective region and country. Resourceful entrepreneurs often have alternative sources of capital lined up before needing it and are in constant communication with potential investors giving them updates on the state of the business.

In terms of type of financing, equity capital in the form of common or preferred stock is the preferred choice of financing for an emerging technology company developing graphene-enabled products. Debt financing can be employed, but this type of funding is discouraged for graphene ventures due to the unpredictable nature of future company cash flows. Debt financings are far better suited toward later stage companies in more stable industry segments that are generating free cash flow. That said, it is quite common to see emerging technology companies using convertible debt – that is, debt that converts into equity at a later stage of development depending on meeting certain criteria as specific in the funding documents – to fund early stage development. A relatively new and quite attractive option for graphene venture is convertible equity. This type of funding is suitable to the risk and development profile of early stage graphene ventures and makes for an attractive alternative source of funding relative to convertible debt. It is important for a company to have a capital structure that is properly aligned and supportive of the business.

The exhibit in the following text illustrates a hypothetical evolution of funding associated with an early stage graphene venture in the current financial environment. Clearly, financial environments will shift over time and it is important for companies to remain opportunistic and take advantage of as many funding channels as possible. The exhibit shows a relationship between the type of funding and the product development cycle – a cycle that typically extends out over a decade or more for an advanced materials company.

As the exhibit shows, graphene ventures are likely to be predominately funded by founders, F&F early on, along with government grant monies. Some companies may attempt an IPO and/or reverse merger for funding if the channel is available. The funding at this stage of evolution fosters important activities related to customer and product development. Success with product and customer development will embolden executives to seek additional funding from deeper pockets (e.g., HNW and Angel investors). Bringing in bigger sums of capital facilitates further product development and the expansion of important operational activities such as sales and marketing. At this time, executives may be successful in securing strategic partnerships that bring in additional capital and assist with customer development. Product development may be accelerated through additional government/research grants. Validation from customers will further embolden executives to seek additional funding, typically from institutional investors that can write big checks. At this juncture in the company's evolution, there is clear market demand for the venture's graphene-enabled technology or application and sales and marketing activity ramp up to support commercialization. Successful product commercialization is likely to lead executives to consider raising more capital through a public offering or a reverse merger. There may be an opportunity for a liquidity event through an acquisition by a larger, more established company. Founders and executives will often seek liquidity for shareholders as a way to enhance the company's attractiveness to investors and provide a means to reap the rewards of their hard work.

The hypothetical financing cycle posits a tight relationship between funding and customer and product development. It is often the case that the relationship breaks down along the way for various reasons that frequently lie outside the control of the company. It is tempting for founders and executives to rush to do an IPO or reverse merger before customers and products have been developed. Such behavior can backfire in the case of advanced materials companies given the relatively long product development lead times. Investors that participate in an IPO or reverse merger that is completed early on while a company is in the early phase of research and development may lose patience as executives struggle to create value. Chief among the reasons for a breakdown in the relationship are shifts in the financial environment and investment climate. The next section of this chapter explores the shifting financial landscape and discusses the implications for companies innovating with graphene.

Many of the companies mentioned in the first section of this chapter are engaged in collaborations with larger, more established companies. Strategic partnerships can be highly effective in accomplishing two important tasks associated with early stage graphene ventures: raising capital and acquiring customers. There is growing interest in graphene for commercial use by companies such as Samsung, Lockheed Martin, IBM, and many others. Such collaborations can, and often do, provide financing and access to customers, which is critical to the success of an early stage technology venture. Such financings will be highlighted and discussed further below.

8.3
Shifting Financial Landscape for Graphene Ventures

The most prevalent sources of funding for seed or early stage graphene ventures over the past decade have been F&F, HNW, Angel, Government, and Strategic. Notably absent from the funding of graphene ventures is venture capital and IPOs and reverse mergers. To be sure, as noted earlier a few graphene start-ups have been able to complete IPOs during the past couple of years, including Haydale PLC, Applied Graphene Materials, and Graphene 3D Labs (see Table "Graphene Company IPOs").

As the exhibit shows, despite the relatively large amount of hype surrounding graphene, there has not been a large demand for graphene-related IPOs to date. There are many reasons for this. For starters, venture capital funds have been increasingly migrating toward social media/software deals and away from funding advanced nanomaterials companies. We will have more to say about this trend.

Graphene company IPOs

Company	Offering date	Amount raised (USD)	Exchange
Applied Graphene Materials	Nov 2013	17 mn	AIM
Cientifica	Oct 2013	389k	AIM
Haydale PLC	Apr 2014	11 mn	AIM
Graphene 3D Labs[a]	Aug 2014	1.5 mn	TSX VE

a) Reverse merger transaction that included private placement of USD6 million shares at USD0.25.
Source: Graphene-info.

The financing environment for early stage graphene ventures has shifted in recent years. In the period leading up to the announcement of the Nobel Prize there was appreciable interest by investors that were piqued by the remarkable properties of graphene and its potential in a wide range of industries. Interest in funding graphene start-ups escalated in 2010, following Geim and Novoselov's Nobel Prize award for the discovery of the 2D nanomaterial. The hype surrounding graphene intensified following the Nobel Prize award. In 2011–2012, graphene came to be known as "the wonder material" and a poster child for the next disruptive technology, a title previously held by its cousin, carbon nanotubes (CNTs). Interest in graphene ventures came from both retail and institutional investors. Retail investors were captivated by media stories portraying graphene as "a wonder material." Institutional investors saw the potential for game-changing applications with graphene.

During the first couple of years following Geim and Novoselov's Nobel Prize award the funding environment for seed and early stage graphene companies was fairly robust. Numerous graphene companies were seeded during this time. Funding for graphene ventures could be completed within a year with nothing more than a CEO, a Scientific Founder, and a business plan. Investor interest was high from both institutional and retail sources of capital. However, the funding environment for graphene ventures began to shift after Bayer MaterialScience (BMS) announced its exit from the CNT business in spring of 2013.[2]

The BMS news cast a cloud over the commercialization prospects of graphene. Institutional investors began to ask some tough questions due to the lack of commercialization success with CNTs. After all, CNTs were discovered in the early 1990s. After two decades, it would seem reasonable to see evidence of adoption of the advanced nanomaterials. Furthermore, if a company like BMS cannot succeed in the nanocarbon business, why should anyone think a start-up could make it in graphene? These relevant questions require an astute response from the executives associated with fledgling graphene ventures (see "Lessons Learned" section).

2) See "Bayer MaterialScience exits carbon nanotube business," at link: http://www.plasticstoday.com/articles/bayer-materialscience-exits-carbon-nanotube-business0508201301.

The wide range of potential applications for graphene along with the fact that there has been relatively little experience innovating with 2D materials poses unique investment challenges. These challenges are all the more acute in a financing environment where investor interest in advanced materials and nanotechnology is overshadowed by the attraction of social media and other software-related investment opportunities today. This subject is discussed further in the next section of this chapter.

Lessons Learned

BMS's exit from the CNT business in May 2013, while casting a cloud over graphene's commercialization potential and leading to a shift in the financing environment for graphene ventures, has been instructive. There are some valuable lessons associated with BMS's experience that graphene ventures can learn.

Over the past two decades, CNT production, with a few exceptions, followed a traditional structure of the materials value chain. There was an implicit assumption that one could simply produce the materials evermore cheaply and others would add value to them along the way. This assumption turned out to be costly. Commercializing advanced nanomaterials requires new tools and processes that enable dispersion for use with other conventional industrial polymers. Producers of nanomaterials are required to work closely with customers and strategic partners to create valued-added materials with desirable properties at a price/performance ratio that is attractive for the customer.

Advanced nanomaterials such as CNTs are inelastic with respect to their demand. The demand for CNTs is tied to their ability to be used, not to their price. No one wants nanocarbon *per se*. What companies seek are better materials – materials improved by the addition of the most active form of carbon available. In an interview with Ross Kozarsky of Lux Research, Peter Kruger, former head of BMS, admitted that the company should have focused more on application development and working with customers than with raw CNT production. This is precisely the path Nanosys CEO Jason Hartlove and his team took that led to successful market penetration with its quantum dot nanomaterials.

Nanosys was long recognized for its deep portfolio of nanotechnology intellectual property. After taking the helm at Nanosys, Hartlove and his team evaluated all of the IP and saw a near-term path to commercialization with the company's proprietary quantum dot nanomaterials. Hartlove's strategy was to create a high valued-added product with its proprietary quantum dots. Within a few of years of Hartlove taking over the helm, Nanosys was able to bring to market a next generation quantum dot display technology called QDEF™.

The QDEF technology produces vibrant displays that are being used in various consumer electronics products today.

The Nanosys model for commercializing advanced nanomaterials stands in contrast to the failed approach by BMS with CNTs. In sum, BMS' failure with CNTs has been an instructive and a valuable lesson in how not to commercialize advanced nanomaterials. Given the lessons being learned, it would be misguidance to extrapolate the failed fortunes of CNTs to graphene, as some investors were prone to do after BMS exited the CNT business. It is the failures of the past that lead to the successes of tomorrow for those who are paying attention. Entrepreneurs innovating with graphene and the investors backing them are advised to take note of the important lessons learned in commercializing advanced nanomaterials.

Start-ups intent on commercializing graphene and other advanced 2D and 3D nanomaterials face many hurdles in the capital markets today. Chief among the hurdles is the relatively long length of time and large amount of capital required to meaningfully create a new market or penetrate an existing market (see the "Nantero Case Study" in the Appendix to better appreciate the time, capital, and effort to bring an innovative nanomaterials product to market). By contrast, early stage social media companies can leverage the hundreds of millions of computer devices connected to the Internet and ramp up in a matter of months or a couple of years.

Witness the ascent social media companies such as Uber that are able raise hundreds of millions and billions of venture capital today. According to *CB Insights*, Uber has raised USD5.5 billion since the company's founding in March 2009. The most recent round of venture financing commanded a USD50 billion PE valuation.[3] Eye-popping numbers such as these would make any venture investor stand up and take notice. By contrast, there are advanced nanomaterials companies that have been plugging away for over a decade and have raised a fraction of what Uber has raised in terms of capital and at valuations that are a small percentage of what Uber is commanding.

To be sure, not all social media and software ventures are akin to Uber. Other so-called early stage "on demand" companies in the United States have raised USD5.9 billion, slightly more than Uber alone. Nonetheless, billions of dollars of early stage venture capital is flowing into the social media segment today due to the favorable business climate and dynamics for such enterprises. This poses a major challenge for graphene start-ups today. There is a large and growing number of social media start-ups vying for capital today that have the potential to leverage the hundreds of millions of electronic devices in the market today and ramp up very quickly.

3) See *CB Insights* blog post at link https://www.cbinsights.com/blog/uber-bigger-entire-on-demand-economy/.

8.4
The Graphene Financing Road Ahead

Due to the success of social media companies such as Uber, Facebook, LinkedIn, Twitter, and others, social media start-ups are commanding the majority of attention of venture investors in Silicon Valley and elsewhere. Entrepreneurs seeking to commercialize graphene or some other advanced nanomaterial must recognize and take into consideration the competition for capital coming from social media ventures – and more generally, software ventures – today. A typical VC or Angel investor is far more interested in funding a social media business that has Uber-like potential than an advanced nanomaterials start-up. VC Steve Blank noted this attraction in an illuminating blog post titled, "Why Facebook is Killing Silicon Valley." In his blog post, Blank wrote:

> *"If investors have a choice of investing in a blockbuster cancer drug that will pay them nothing for fifteen years or a social media application that can go big in a few years, which do you think they're going to pick? If you're a VC firm, you're phasing out your life science division. As investors funding clean tech watch the Chinese dump cheap solar cells in the U.S. and put U.S. startups out of business, do you think they're going to continue to fund solar? And as Clean Tech VC's have painfully learned, trying to scale Clean Tech past demonstration plants to industrial scale takes capital and time past the resources of venture capital. A new car company? It takes at least a decade and needs at least a billion dollars. Compared to IOS/Android apps, all that other stuff is hard and the returns take forever. Instead, the investor money is moving to social media. Because of the size of the market and the nature of the applications, the returns are quick – and huge. New VC's, focused on both the early and late stage of social media have transformed the VC landscape."*[4)]

Blank's assessment of the investing landscape in Silicon Valley is sobering. Software has been eating the world and venture capital deal flow in Silicon Valley and elsewhere has been increasingly concentrated in social media for all the reasons noted in his blog post. That said, it remains to be seen if these trends will continue in the years ahead. Investment fads come and go, as all experienced investors know. Social media is hot now, but may not always be hot. There are already signs that the ebullience in social media investors is leading to over-inflated valuations and resource constraints that could turn the tide of increasing concentration in the software segment.

As Blank notes, the unwritten manifesto for Silicon Valley VCs for decades has been: "*We choose to invest in ideas, not because they are easy, but because they are hard, because that goal will serve to organize and measure the best of our energies and skills, because that challenge is one that we are willing to accept, one we are*

4) See Steve Blank blog post, "Why Facebook is Killing Silicon Valley," at link http://steveblank.com/2012/05/21/why-facebook-is-killing-silicon-valley/.

unwilling to postpone, and one which we intend to win." Graphene has tremendous commercial potential in the marketplace. There is hope that 1-day VCs in Silicon Valley and elsewhere see attractive investment opportunities once again in funding start-ups and emerging companies seeking to innovate with advanced nanomaterials.

Entrepreneurs seeking to start-up graphene ventures need to be especially creative and resourceful in how they go about funding and operating their companies in an environment where software is eating the world. It will be incumbent upon all emerging graphene companies to tap into as many investment channels as possible and formulate strategies that foster accelerated product development times and strategic partnerships that promote rapid customer development as well as access to financial capital.

There are signs today that governments around the world are increasingly committed to fostering innovation associated with advanced nanomaterials such as graphene. The Graphene Flagship in Europe is a prime example. Government funding initiatives announced in recent years have been geared toward basic scientific research – research that is essential and important, as basic scientific research has the potential to spill over into the development of commercial applications. There is heightened interest among various governments around the world to fund the development of graphene-enabled applications for various defense and military applications. Entrepreneurs tapping into government funding sources for such applications are always encouraged to seek to leverage any success in that segment into the enterprise and consumer segments. After all, there is no substitute for customer financing. Successful and sustainable businesses, graphene and otherwise, are funded by customers that benefit from using the product or application.

To the extent government funding is available, we encourage graphene start-up companies and emerging graphene ventures to continue to monitor the investment landscape carefully. There have been some constructive and noteworthy developments in US capital markets that seek to foster greater access to capital for emerging technology companies. In March 2015, the Securities and Exchange Commission (SEC) announced new rules to provide investors with more investment choices.

The new rules update and expand Regulation A, an existing exemption from registration for smaller issuers of securities. The rules are mandated by Title IV of the Jumpstart Our Business Startups (JOBS) Act. The updated exemption will enable smaller companies to offer and sell up to USD50 million of securities in a 12-month period, subject to eligibility, disclosure, and reporting requirements. SEC Chair Mary Jo White noted that it is important for the Commission to continue to look for ways to facilitate capital raising by smaller companies.[5)]

With Title IV of the 2012 JOBS Act, the fastest growing private companies in America can conduct IPOs and raise up to USD50 million from any investor.

5) See press release, "SEC Adopts Rules to Facilitate Smaller Companies' Access to Capital," at link, http://www.sec.gov/news/pressrelease/2015-49.html.

Historically, investing in start-ups and small businesses in the United States was reserved for just accredited investors, or just the wealthiest 2% of America. Now, after 3 years of anticipation, investing in pre-IPO companies will be opened up to the other 98%. The new SEC legislation is a breath of fresh air and a welcome development for emerging graphene ventures. The new SEC legislation could provide further impetus to the crowdfunding platforms that have emerged in recent years that are friendly to start-up companies. It remains to be seen how entrepreneurs take advantage of the new rules and how receptive investors are in funding early stage advanced nanomaterials companies in US capital markets.

Meanwhile, emerging graphene companies in the United Kingdom and Europe will continue to raise funds through small capitalization listings on the AIM. The AIM has been friendly to entrepreneurs raising capital for graphene-enabled innovations and is likely to remain so in the future. Of course, graphene entrepreneurs always need to be mindful that the window for small cap IPOs opens and shuts depending on gyrations in the financial markets and the appetite for such deals among investors.

Summary

The discovery of graphene in 2004 opened the door to a wide range of commercialization opportunities. The commercialization of graphene requires funding. The funding of graphene ventures is more of an art than a science. As noted earlier, there are numerous sources of capital available to seed or fund an early stage graphene venture. A resourceful executive will be as opportunistic as possible and try to tap into as many channels of funding as possible to produce and/or commercialize graphene-enabled products and applications. Government funding remains an attractive source of capital for fledgling graphene ventures, particularly in the United Kingdom, Europe, and Asia. The European Union has ploughed over USD1 billion into start-up companies during the past year in an attempt to accelerate the high-tech economy and stimulate economic dynamism. The EU program is targeted to address the chronic complaint from entrepreneurs that funding is far more difficult to raise in Europe than elsewhere. In North America, HNW and Angel funding are likely to continue to be prominent sources of capital for early stage graphene ventures. Globally, crowdfunding is a relatively new channel that could play a bigger role in funding graphene ventures in the future.

The past decade has seen a migration of venture capital funding away from advanced nanomaterials companies and toward social media/software deals. It is unclear if this migration will continue. Nevertheless, the relatively long development cycles of advanced nanomaterials-enabled products and applications present funding challenges as many investors' time horizons have appeared to shorten with the rise of high-frequency trading, hedge funds, and the like. IPOs and reverse mergers are attractive financing options for graphene ventures today, but investor demand for such IPOs with the bigger, more established banks has not been very enthusiastic to date. Regulatory changes in the United States are encouraging and may foster greater access to capital for graphene ventures in the months ahead.

Today, there is capital available to fund seed and early stage graphene ventures. Experience suggests being artful in funding such ventures, pursuing diligently as many sources of capital as possible. Importantly, while a founder or executive may have a well thought out plan of how to finance a graphene venture, it is important to remain opportunistic and take what the market will give you. The fact of the matter is that the volatility of capital markets is unpredictable and what a company can do funding-wise from week to week, month to month, quarter to quarter and year to year can shift significantly and sometimes drastically. It is interesting to note how many founders and executives, when faced with a funding decision that will result in having a smaller share of something potentially big or a large share of something that will likely be small, will choose the latter option. Experience has shown that pride and greed can be major impediments to a successful venture. Founders and executives of graphene ventures are strongly advised to be mindful of the pride and greed factors as they go forward. Lastly, there is a strong case to be made for graphene ventures collaborating with larger, established companies where such strategic alignments make good business sense. Well-crafted strategic partnerships help foster the commercialization of innovative graphene-enabled products to the benefit of all major constituents including, most importantly, the customer.

Appendix

Nantero Case Study – The Funding and Evolution of a Nanomaterials Start-up

This case study explores the founding and funding of Nantero, Inc. Nantero is pioneering the use of CNTs in memory devices. While silicon may continue to be the dominant material for the production of microprocessors and other types of semiconductors over the next 5 years or so, scientists and researchers are exploring the use of other materials – materials that are highly scalable down to 1 or 2 nm and that would keep Moore's law chugging along for the foreseeable future. One material that appears promising and has received a great deal of attention over the past decade is CNTs.

CNTs are incredibly strong elastic cylinders of carbon atoms that bear a striking resemblance to a tube of rolled-up chicken wire. One nanotube is just 1/50 000th the diameter of a human hair. CNTs possess unique structural and electrical properties. They are 117 times stronger than steel, half the density of aluminum, and have thermal and electrical conductivity properties that make the material useful for a variety of applications in a wide range of industries from aerospace and transportation to energy and electronics. Since their discovery in 1991, CNTs have been a focus of intense research in the United States and overseas, being viewed as a potential replacement for silicon in semiconductors in the future. The high electrical conductivity, thermal conductivity, and tensile strength of CNTs make them highly attractive for electronic device applications. These properties enable performance breakthroughs both through incorporation into existing semiconductor products and in the development of next-generation products.

The Founding of Nantero

One of the companies pioneering the use of CNTs in computer chips is Nantero. Nantero was founded in Woburn, MA, in 2001 by Dr Thomas Rueckes, Greg Schmergel, and Dr Brent Segal. Dr Rueckes is the inventor of the innovative nanoelectromechanical NRAM™ design and the company's Chief Technology Officer. Greg Schmergel is a successful entrepreneur and the company's CEO and Dr Segal was the COO. Rueckes, Schmergel, and Segal assembled a complete scientific team consisting of specialists in multiple disciplines in the fields of nanotechnology and semiconductors to develop Nantero's nanotechnology.

Nantero was launched with the objective of developing and commercializing Dr Rueckes' NRAM invention – a high-density nonvolatile random access memory using CNTs. Nantero's founders envisioned NRAM as a universal memory device that would replace all existing forms of memory, such as DRAM, SRAM, and flash memory. Additionally, the company's founders sought to develop a proprietary way to manufacture the nanotechnology memory chip and integrate it with standard semiconductor processes. Development of such processes was seen as a way to advance Nantero's CNT memory device into the market more quickly.

At the time of founding, Nantero envisioned a market opportunity for its NRAM chip in excess of USD70 billion per year. NRAM would enable instant-on computers and replace DRAM and flash memory in devices such as smartphones, tablets, laptops, MP3 players, digital cameras, and enterprise systems of all kinds. There are also other applications in the networking segment. Nantero's business model is based on licensing the company's proprietary CNT technology and processes to manufacturers for use in various applications. As such, Nantero is a next generation fabless semiconductor company, one focused primarily on IP licensing, similar to ARM Holdings.

Series A: Financing Round

In October 2001, Nantero announced its first round of investment to develop a nanotube-based universal memory. The amount raised in the Series A round was USD6 million. The proceeds of the capital raised were to be used to develop Nantero's core NRAM technology and processes. Investors in the Series A round included venture firms Draper Fisher Jurvetson, Stata Venture Partners, and Harris & Harris Group. Other investors included Alex d'Arbeloff, Chairman of MIT and founder of Teradyne, who also joined Nantero's Board of Directors.

It should be noted that the financial environment in 2001 was not particularly conducive to raising significant sums of capital for emerging technology companies – especially companies pioneering the development of innovative nanotechnologies. In the year leading up to the completion of Nantero's Series A round, the NASDAQ stock index – a bellwether for emerging technology stocks – had declined sharply, driven by a collapse of valuations in the previously high-flying dot com segment.

Post-Series A: Funding Evolution

In the spring of 2003, Nantero announced that it had achieved a milestone that involved creating an array of 10 billion suspended nanotube junctions on a single silicon wafer. This development showed that nanotubes could reliably be positioned in large arrays and was able to scale to make even larger arrays. Nantero's process also resulted in substantial redundancy for the memory, because each memory bit depends not on one single nanotube, but upon a large number of nanotubes that resemble a fabric. The highly conductive single-layer nanotube fabrics were viewed as having a wide range of applications beyond memory chips, including for transistors, interconnects, and sensors. Creating this enormous array of suspended nanotubes using standard semiconductor processes brought Nantero closer to its end goal of mass-producing NRAM chips.

Nantero's innovative design for NRAM involved the use of suspended nanotube junctions as memory bits, with the "up" position representing bit zero and the "down" position, representing bit one. Bits are switched between states through the application of electrical fields. The wafer was produced using only standard semiconductor processes, maximizing compatibility with existing semiconductor factories. Nantero's proprietary method for achieving this result involved depositing a very thin layer of CNTs over the entire surface of the wafer, and then using lithography and etching to remove the nanotubes that are not in the correct position to serve as elements in the array.

During 2003, Nantero added Dr Mohan Rao to its Scientific Advisory Board. Dr Rao was one of the world's leading VLSI chip designers, having served previously as Senior Vice President, Semiconductor Group, at Texas Instruments. He held over 100 patents worldwide on various aspects of memory, including SRAM, DRAM, and system-on-chip.

Series B: Financing Round

In September 2003, Nantero announced the completion of a Series B round of financing. The company raised USD10.5 million to further advance the development of its innovative NRAM technology. The Series B round was led by Charles Rivers Ventures, a venture firm with over three decades of experience in high technology. Bruce Sachs and Bill Tai, both Partners at Charles River Ventures, joined Nantero's Board of Directors. Returning existing institutional investors included Draper Fisher Jurvetson, Stata Venture Partners, and Harris & Harris Group.

Post-Series B: Funding Evolution

Around the same time as its Series B funding round, Nantero announced collaboration with ASML, the leading global provider of lithography systems for the semiconductor industry headquartered in Veldhoven, the Netherlands. Nantero had developed a completely CMOS-compatible manufacturing process for NRAM. The joint work with ASML was geared toward proving that Nantero's proprietary process was achievable using popular lithography equipment. The collaboration demonstrated that ASML's equipment is fully capable of handling nanotubes using Nantero protocols and of carrying out Nantero's new manufacturing steps without any modifications. This was a significant milestone for the company and set it on course to advance commercialization of its CNT memory devices.

In the spring of 2004, Nantero announced that OB Bilous joined the Company's Board of Directors. At the time Mr Bilous was Chairman of the Board of International SEMATECH and held 15 patents and had published numerous articles and presented in many conferences and technical meetings in the area of semiconductor technology. Mr Bilous was previously VP of Worldwide Manufacturing for IBM in microelectronics.

In the summer of 2004, Nantero announced that it had been issued a seminal patent covering CNT films and fabrics by the US Patent and Trademark Office. The patent relates to a CNT film comprising a conductive fabric of CNTs deposited on a surface. The CNT film is highly useful in a variety of applications, including in the semiconductor industry wherein the CNT film is deposited on a silicon substrate. The CNT film is a major innovation enabling cost-effective, high-volume production of CNT-based devices and other products. With the addition of the new patent, Nantero had grown its patent portfolio to 10 granted US patents and a pipeline of over 40 additional patents pending.

Also in the summer of 2004, Nantero announced a CNT development project with LSI Logic to develop semiconductor process technology, expediting the effective use of CNTs in CMOS fabrication. During the same time, the company also announced a joint evaluation of CNT-based electronics with BAE Systems to evaluate the potential to develop CNT-based electronic devices for use in advanced defense and aerospace systems. The project involved research and development of a variety of next-generation electronic devices that can be built leveraging the unique properties of CNTs and using Nantero's proprietary methods and processes for the design and manufacture of nanotube-based electronics.

In the fall of 2004, Nantero announced that it had been awarded USD4.5 million to develop CNT-based radiation-hard non-volatile RAM with CNI and Case-Southwest Missouri State University. The radiation-resistant nanotechnology was targeted for use in space for US defense purposes. In early 2005, Nantero announced it was actively seeking manufacturing partners in Europe and Asia to further advance the commercialization of its NRAM in the consumer electronics segment. Germany, France, Italy, the Netherlands, Japan, and Korea were among the countries earmarked for possible licensing of the NRAM technology.

Series C: Financing Round

Nantero announced the completion of a USD15 million Series C round of funding in the first quarter of 2005. Globespan Capital Partners was the lead investor in the round and Ullas Naik, a Managing Director with Globespan Capital Partners based in their Boston office, joined Nantero's Board of Directors. Returning existing institutional investors included Charles River Ventures, Draper Fisher Jurvetson, Stata Venture Partners, and Harris & Harris Group.

Post-Series C: Funding Evolution

In spring of 2006, Nantero announced it had fabricated and successfully tested a 22-nm NRAM memory switch. The switch demonstrated that NRAM is scalable to numerous process technology nodes over several decades. Nantero's NRAM switches were tested by writing and reading data using 3 ns cycle times, giving it the potential to match the fastest memories in production at the time. The NRAM switches were fabricated using the company's proprietary CNT fabric. The results demonstrated that NRAM can be the standalone and embedded memory of choice as the technology combines the nonvolatility of flash with the speed of SRAM and the density of DRAM. The tests showed that NRAM can be scaled for many future generations and we believe the scaling will continue down to below the 5 nm technology node. Also in the spring of 2006, Nantero announced collaboration with ON Semiconductor to jointly develop CNT technology, continuing ongoing work to integrate CNTs in CMOS fabrication.

In fall 2006, Nantero announced a major corporate milestone: that it has resolved all of the major obstacles that had been preventing CNTs from being used in mass production in semiconductor fabs. This was a significant development for the company. Nanotubes were widely acknowledged to hold great promise for the future of semiconductors, but most experts had predicted it would take a decade or two before they would become a viable material. This was due to several historic obstacles that prevented their use, including a previous inability to position them reliably across entire silicon wafers and contamination previously mixed with the nanotubes that made the nanotube material incompatible with semiconductor fabs.

Nantero has developed a method for positioning CNTs reliably on a large scale by treating them as a fabric that can be deposited using methods such as spin-coating, and then patterned using lithography and etching, all common CMOS processes present in every semiconductor fab. Nantero also developed a method for purifying CNTs to the standards required for use in a production semiconductor fab, which means consistently containing less than 25 parts per billion of any metal contamination. With these innovations, Nantero became the first company in the world to introduce and use CNTs in mass production semiconductor fabs.

The summer of 2007 saw Nantero providing license rights for biomedical sensors to Alpha Szenszor, Inc., a newly formed company also based in Woburn, Massachusetts. Alpha Szenszor was founded to develop a suite of sensor products in the medical field, including portable and cost-effective detectors for infectious diseases such as HIV. Alpha Szenszor's co-founders include Steve Lerner, an industry veteran with over 28 years of product development and industrialization experience. Additionally during this time, Nantero announced it was working with Hewlett-Packard (HP) to explore the use of HP inkjet technology and the company's CNT formulation to create flexible electronics products and develop low cost printable memory applications. Nantero used HP's thermal inkjet pico-fluidic system (TIPS) research and development tool to evaluate the company's inkjet technology for printable memory applications that can be used in a wide range of applications including low cost RFID tags.

During summer of 2008, Nantero announced collaboration with SVTC Technologies to accelerate the commercialization of nanotube-based electronics products. The collaboration was part of SVTC's broader mission to enable commercialization of new process and device developments in the semiconductor, MEMS, and related nanotechnology domains with support for a direct path between the work completed in SVTC's facilities to high-volume manufacturing. Together, Nantero and SVTC offered CNT device development capabilities for customers targeting a wide range of applications including photovoltaics (solar cells), LEDs, sensors, MEMS, and other semiconductor-based devices.

Also in 2008 Nantero announced a significant milestone with the acquisition of its government business unit by Lockheed Martin. Additionally, Lockheed Martin entered an exclusive license arrangement with Nantero for government-specific applications of Nantero's extensive intellectual property portfolio. Approximately 30 employees, including Nantero's co-founder and COO Dr Brent Segal joined Lockheed Martin as part of the purchase of Nantero's government business unit. Lockheed Martin's Advanced Technology Center, a unit of Lockheed Martin Space Systems Company, would manage the Nantero unit going forward. At the time, Lockheed Martin was a leader in the research, development, and application of nanotechnology to future military and intelligence applications. Deal terms were undisclosed. The year 2008 also saw Nantero ranked Number 54 in *Inc. Magazine's* "Annual List of America's 500 Fastest-Growing Private Companies, with three-year sales Growth of 2833%. Along with the ranking of 54 in the Inc 500, Nantero was #2 in the computer and electronics category and #2 for Boston area companies.

In late 2009, Lockheed Martin announced that it tested a radiation-resistant version of Nantero NRAM devices on a NASA Shuttle mission. NASA incorporated the NRAM technology into special autonomous testing configurations installed into a carrier at the aft end of the payload bay. It was launched into space as part of STS-125, the May 2009 mission of the Space Shuttle Atlantis that successfully serviced the Hubble Space Telescope. The experiment was a proof-of-concept that enabled the testing of launch and re-entry survivability, as well as basic functionality of the CNT memory in orbit throughout the shuttle mission. The NRAM devices were early prototype parts, and performed the same before, during, and after completion of the mission. This mission represents an important first step in the development of high-density, non-volatile, carbon-nanotube-based memories for spaceflight applications.

Series D: Financing Round

In 2013, Nantero announced the closing of a Series D round of funding that raised more than USD15 million. The round also included existing investors Charles River Ventures, Draper Fisher Jurvetson, Globespan Capital Partners, Stata Venture Partners, and Harris & Harris Group. The round also included new strategic corporate investors. One of those investors is Schlumberger, the world's leading supplier of technology, integrated project management and information solutions to customers working in the oil and gas industry worldwide. Additionally, Michael Raam was added to the Advisory Board. Mr Raam was most recently the CEO of SandForce, Inc., the successful SSD controller company and then VP/GM of the Flash Components Division for LSI after LSI acquired SandForce.

Also, near the end of 2009, Nantero announced that Dr Tsugio Makimoto, a Japanese semiconductor industry pioneer, joined its Advisory Board. Previously, Dr Makimoto was a Corporate Advisor of Sony Corporation in charge of semiconductor technology. It was noted that Dr Makimoto's deep knowledge and experience in the semiconductor industry, both in Japan and globally, would assist in helping the company continue to add new customers in Asia.

Post-Series D: Funding Evolution

In 2014, a team of researchers in Japan independently verified that Nantero's NRAM, had excellent properties and could be used for various applications ranging from main memory to storage. Ken Takeuchi, professor at the Department of Electrical, Electronic and Communication Engineering, Faculty of Science and Engineering of Chuo University, led the Japanese team that conducted the NRAM research. The details of the new technology were announced in a lecture titled "23% Faster Program and 40% Energy Reduction of CNT Non-volatile Memory with Over 10^{11}Endurance" (lecture number: T11-3) at 2014 Symposia on VLSI Technology and Circuits in Honolulu, Hawaii. D. Takeuchi was previously leading Toshiba's NAND flash memory circuit design for 14 years.

Series E: Financing Round

In spring of 2015, Nantero announced the completion of a USD31.5 million Series E financing round. The round included new investors and participation from existing investors Charles River Ventures, Draper Fisher Jurvetson, Globespan Capital Partners, and Harris & Harris Group. The company noted that the round was substantially oversubscribed. Nantero noted that it intended to use its new funding to continue the acceleration of NRAM as the leading next-generation memory for both storage class memory and as a replacement for flash and DRAM. In addition to the funding, Nantero noted that it added two new company advisors. Among the new advisors were Dr Stefan Lai, a former Intel senior executive who co-invented the EPROM tunnel oxide (ETOX) flash memory cell and led the company's phase change memory (PCM) team, and Dr Yaw Wen Hu, a former Executive VP and current Board member of Inotera Memories, where he oversaw new DRAM technology transfer and development of Wafer Level Packaging. Before that, Dr Hu was an Executive VP and Chief Operating Officer for Silicon Storage Technology (SST), where he was responsible for SuperFlash technology development, working with a team from a brand new memory cell concept to high volume product shipment and its establishment as the choice of technology for Embedded Flash applications.

In summer 2015, Nantero announced that arrays of its new generation of super-fast, high-density memory (NRAM) were independently tested by Chuo University and the results showing excellent performance and reliability will be presented in a technical paper at the 2015 International Conference on Solid State Devices and Materials. Nantero announced that former TSMC executive Dr Shang-Yi Chiang had joined the company's advisory board. With more than 40 years experience in the semiconductor industry, including as co-COO and EVP at TSMC, Dr Chiang has contributed to the research and development of CMOS, NMOS, Bipolar, DMOS, SOS, SOI, GaAs lasers, LED, E-Beam lithography, and silicon solar cells. In addition, Nantero also announced the addition of Lee Cleveland to the executive management team. Formerly in charge of flash design at Spansion and AMD, Mr Cleveland is VP of Design and responsible for leading Nantero's complete chip design team. The company also announced the appointment of renowned memory industry executive Ed Doller to its Advisory Board. Mr Doller was previously VP & Chief Strategist of the NAND Solutions Group at Micron, where he also served as VP & GM Enterprise Storage and as VP & Chief Memory Systems Architect.

Summary

The development of Nantero spans 15 years of intensive research and development in the area of advanced nanomaterials. With one CNT being just 1/50 000th the diameter of a human hair, these tiny cylinders are far stronger than steel, half the density of aluminum, and have better thermal and electrical conductivity properties than any other material scientists are aware of today. As a pioneer in nanotechnology, Nantero is the first company to develop semiconductor products using this material in production of CMOS fabs. Nantero's innovative NRAM technology has several characteristics that make it attractive as a next-generation technology for standalone and embedded usage. These include:

- CMOS compatible: Works in standard CMOS fabs with no new equipment needed
- Limitless scalability: Designed to scale below 5 nm in the future
- High endurance: Proven to operate for orders of magnitude more cycles than flash
- Faster Read and Write: Same as DRAM, 100s of times faster than NAND
- High reliability: will retain memory for >1000 years at 85 °C or more than 10 years at 300 °C

- Low power: Zero in standby mode, 160x lower write energy per bit than NAND
- Low cost: Simple structure, can be 3D multi-layer and multi-level cell (MLC)

One of the key advantages of Nantero's NRAM technology is the ability to consolidate DRAM with flash memory, since NRAM is both as fast as DRAM and nonvolatile like flash. This ability will undoubtedly be attractive to manufacturers seeking to create smaller, more powerful devices in the future. Potential future applications of the NRAM technology include: Virtual Screens, Next-Generation Enterprise Systems, Rolled-up Tablets, Instant-On Laptops, 3D Video Phones, and other products needing huge amounts of fast memory, targeting a wide range of markets such as consumer electronics, mobile computing, wearables, Internet of Things, enterprise storage, government/military, space, and automotive.

Nantero has raised over USD87 million of capital over the past 15 years and earned over USD70 million in revenue. The company's nanotechnology is under development in multiple world-class manufacturing facilities and the company has more than a dozen major corporate partners actively working on NRAM. Sample NRAMs have demonstrated production-ready yields (i.e., >99.999%). Performance for both standalone and embedded memory applications are superior to anything in the market. Nantero is currently sampling memory test chips to customers at a time when there is a large demand for new high-density standalone memory and for high reliability, scalable embedded memory.

Nantero has the potential to scale its business significantly beyond NRAM through licensing of its extensive CNT patent portfolio for other applications in a wide variety of sectors. The company's business model is similar to that of ARM Holdings, PLC, a peer that has risen to prominence over the years and today is generating over a half a billion of revenue per annum. There is considerable opportunity for Nantero and its CNT technology in the USD330 billion-plus semiconductor industry as chips scale further down into the quantum realm with advanced nanomaterials that have properties that extend beyond silicon. The age of CNT electronics has begun.

From this case study we see the time, effort, and patience required to commercialize a transformative deep science-enabled technology such as NRAM relative to a social media application circa 2012–2014. Nantero's path to commercialization extends over a period of 15 years. The company has progressed with the development of its innovative nanotechnology without having access to the public capital markets. While a significant amount of private investment capital has been invested in Nantero – over USD87 million to date – the company would have no doubt benefitted from having access to the public capital markets to accelerate the commercialization of its technology. As such, Nantero was unable to tap the public markets for growth capital as many of its silicon brethren did in the decades of the 1970s, 1980s, and 1990s. Whether the company will be able to tap the public capital markets for funding in the future remains to be seen.

Index

Graphene Technology: From Laboratory to Fabrication,
First Edition. Edited by Soroush Nazarpour and Stephen R. Waite.